Lecture Notes of the Institute for Computer Sciences, Social Informatics and Telecommunications Engineering 408

More information about this series at https://link.springer.com/bookseries/8197

Phan Cong Vinh · Nguyen Huu Nhan (Eds.)

Nature of Computation and Communication

7th EAI International Conference, ICTCC 2021
Virtual Event, October 28–29, 2021
Proceedings

 Springer

Editors
Phan Cong Vinh (ID)
Nguyen Tat Thanh University
Ho Chi Minh City, Vietnam

Nguyen Huu Nhan
Nguyen Tat Thanh University
Ho Chi Minh City, Vietnam

ISSN 1867-8211   ISSN 1867-822X (electronic)
Lecture Notes of the Institute for Computer Sciences, Social Informatics
and Telecommunications Engineering
ISBN 978-3-030-92941-1   ISBN 978-3-030-92942-8 (eBook)
https://doi.org/10.1007/978-3-030-92942-8

This Springer imprint is published by the registered company Springer Nature Switzerland AG
The registered company address is: Gewerbestrasse 11, 6330 Cham, Switzerland

Preface

ICTCC 2021 (the 7th EAI International Conference on Nature of Computation and Communication) was held during October 28–29, 2021, in cyberspace due to the travel restrictions caused by the worldwide COVID-19 pandemic. The aim of the ICTCC conference series is to provide an internationally respected forum for scientific research in the technologies and applications of natural computing and communication. This conference provides an excellent opportunity for researchers to discuss modern approaches and techniques for natural computing systems and their applications. The proceedings of ICTCC 2021 are published by Springer in the series Lecture Notes of the Institute for Computer Sciences, Social Informatics and Telecommunications Engineering (LNICST; indexed by DBLP, EI, Google Scholar, Scopus, and Thomson ISI).

For this seventh edition of ICTCC, repeating the success of the previous year, the Program Committee received submissions from authors in six countries and each paper was reviewed by at least three expert reviewers. We chose 17 papers after intensive discussions held among the Program Committee members. We really appreciate the excellent reviews and lively discussions of the Program Committee members and external reviewers in the review process. This year we had two workshops, one on Applied Mathematics on Sciences (AMS) chaired by Nguyen Huu Nhan from Nguyen Tat Thanh University (NTTU) and one on Big Data and Machine Learning (BD-ML) chaired by Do Van Thanh from NTTU. Moreover, we had three prominent invited speakers, Francois Siewe from De Montfort University in the UK, Kurt Geihs from the University of Kessel in Germany, and Hafiz Mahfooz Ul Haque from the University of Lahore in Pakistan.

ICTCC 2021 was jointly organized by the European Alliance for Innovation (EAI), Ho Chi Minh City Open University (OU), and NTTU. This conference could not have been organized without the strong support of the staff members of the three organizations. We would especially like to thank Imrich Chlamtac (University of Trento), Aleksandra Śledziejowska (EAI), and Martin Karbovanec (EAI) for their great help in organizing the conference. We also appreciate the gentle guidance and help of Nguyen Minh Ha, Rector of OU.

October 2021

Phan Cong Vinh
Nguyen Huu Nhan

Organization

Steering Committee

Imrich Chlamtac (Chair)   University of Trento, Italy
Phan Cong Vinh   Nguyen Tat Thanh University, Vietnam

Organizing Committee

Honorary General Chair

Nguyen Minh Ha   Ho Chi Minh City Open University, Vietnam

General Chair

Phan Cong Vinh   Nguyen Tat Thanh University, Vietnam

Program Chair

Waralak V. Siricharoen   Silpakorn University, Thailand

Workshop Chair

Hafiz Mahfooz Ul Haque   University of Lahore, Pakistan

Publicity Chair

Nguyen Kim Quoc   Nguyen Tat Thanh University, Vietnam

Publication Chair

Phan Cong Vinh   Nguyen Tat Thanh University, Vietnam

Sponsorship and Exhibits Chair

Bach Long Giang   Nguyen Tat Thanh University, Vietnam

Local Arrangement Chair

Le Xuan Truong   Ho Chi Minh City Open University, Vietnam

Web Chair

Nguyen Van Han   Nguyen Tat Thanh University, Vietnam

Technical Program Committee

Abdur Rakib	University of the West of England, UK
Amando P. Singun Jr.	University of Technology and Applied Sciences, Oman
Bui Cong Giao	Saigon University, Vietnam
Chernyi Sergei	Admiral Makarov State University of Maritime and Inland Shipping, Russia
Chien-Chih Yu	National Chengchi University, Taiwan
David Sundaram	University of Auckland, New Zealand
Do Tri Nhut	Thu Dau Mot University, Vietnam
Gabrielle Peko	University of Auckland, New Zealand
Giacomo Cabri	University of Modena and Reggio Emilia, Italy
Huynh Xuan Hiep	Can Tho University, Vietnam
Ijaz Uddin	City University of Science and Information Technology, Peshawar, Pakistan
Issam Damaj	The American University of Kuwait, Kuwait
Jinfeng Li	Imperial College London, UK
Krishna Asawa	Jaypee Institute of Information Technology, India
Kurt Geihs	University of Kassel, Germany
Le Hong Anh	University of Mining and Geology, Vietnam
Manisha Chawla	Google, India
Muhammad Athar Javed Sethi	University of Engineering and Technology, Peshawar, Pakistan
Nguyen Hoang Thuan	RMIT University Vietnam, Vietnam
Nguyen Manh Duc	University of Ulsan, South Korea
Nguyen Thanh Binh	Ho Chi Minh City University of Technology, Vietnam
Nguyen Thanh Hai	Can Tho University, Vietnam
Pham Quoc Cuong	Ho Chi Minh City University of Technology, Vietnam
Shahzad Aahraf	Hohai University, China
Tran Huu Tam	University of Kassel, Germany
Tran Vinh Phuoc	Ho Chi Minh City Open University, Vietnam
Vu Tuan Anh	Industrial University of Ho Chi Minh City, Vietnam
Zhu Huibiao	East China Normal University, China

Workshop Chairs

Nguyen Huu Nhan	Nguyen Tat Thanh University, Vietnam
Do Van Thanh	Nguyen Tat Thanh University, Vietnam

Contents

Vietnamese Export Growth Prediction Applying MIDAS and MF-VAR on Mixed-Frequency Data

Nguyen Thi Hien[1], Hoang Anh Tuan[2], Dinh Thi Ha[3(✉)],
Le Mai Trang[2], Tran Kim Anh[2], and Dao The Son[2]

[1] Mathematics Department, Thuongmai University, Hanoi, Vietnam
hiennguyen@tmu.edu.vn
[2] Economics Department, Thuongmai University, Hanoi, Vietnam
{hoanganhtuan, lmtrang2000, trankimanh,
daotheson}@tmu.edu.vn
[3] Informatics Department, Thuongmai University, Hanoi, Vietnam
dinhha@tmu.edu.vn

Abstract. Import and export growth forecasting is always a concern for researchers as well as policymakers in any country. However, forecasting this index through methods that employ data sets with the same frequency does not reflect reality due to the fact that the economic and financial indicators have different published frequencies. Therefore, between two data periods, some important information affecting the target variable maybe not be included in the model, leading to the limited accuracy of the forecast. The approach of data analysis with multiple frequencies has gained a lot of interest recently in order to overcome data restrictions and increase forecasting performance. In this article, we study and apply a number of models for mixed-frequency data such as MF-VAR and MIDAS to predict Vietnam's exports based on collected data sets in the period from 2006 to 2020. The findings show that the MF-VAR model has high forecasting error and is not suitable for predicting the export growth of Vietnam, while the MIDAS model gives good prediction results on the same data set. In addition, the prediction results also reveal that the MIDAS model is effective for short-term forecasting. This result is quite similar to that of several previously published studies. The findings of this research also open a promising direction in studying and applying the mixed frequency data models to forecast export growth as well as other economic indicators .

Keywords: Export growth · Forecasting · Mixed-frequency data · MF-VAR · MIDAS

1 Introduction

Trade is a key component of GDP for almost any country. That is particularly true for an export-oriented economy like Vietnam. Trade as a percentage of GDP for Vietnam has increased steadily for the last four decades, starting with 23.3% in 1986 to 209.3% in 2020. Last year, the export of goods and services from Vietnam reached 287.8

P. Cong Vinh and N. Huu Nhan (Eds.): ICTCC 2021, LNICST 408, pp. 1–19, 2021.
https://doi.org/10.1007/978-3-030-92942-8_1

billion USD or the equivalent of 106.1% GDP, while that number for import is 279.8 billion USD or 103.2% GDP (Nguyen et al. 2020). This makes trade an essential part of GDP forecasting for the purpose of macroeconomic policies.

There are numerous swings that accompany the growing level of economic openness and integration. According to trade reports, Vietnam's situation of exports and imports can face uncertainties from many domestic and international factors. Therefore, an accurate forecast of export growth will help policymakers build national economic growth scenarios, as well as help countries be more proactive in their foreign exchange reserves. Besides, the correct and timely forecast of export growth also helps researchers and policymakers to forecast shocks to the economy, because exports have a great influence on the business cycle.

For any macroeconomic forecast, there is a need to have the most precise and most timely estimate possible. This goal does have a common challenge from macroeconomic data when reports contain data with different frequencies. They are not officially made available at the same time, or in many cases, there can be lags in reporting.

This study aims to investigate the performance of Mixed-Frequency models on forecasting export growth in Vietnam. These models create possibilities to utilize both low-frequency and high-frequency data, to combine lag reports with real-time data to give accuracy and timely forecast for Vietnamese export MIDAS and MF-VAR stay in the focusing point, and its parameters are modified to capture the fluctuation of data and to improve forecasting accuracy. The study collects data from different public sources on the supply and demand sides of exports. Data is then used to test the accuracy of different mixed-frequency forecasting models by comparing predicted values with real export growths. This experimental work on export data can serve as a preliminary study for further applications on other essential macroeconomic indicators for the purpose of monitoring and making policies.

The remainder of this paper is structured as follows: Sect. 2 gives an overview of the literature on export forecasting techniques with both types of models, using the same frequency and mixed-frequency data. Section 3 describes the data collection and preprocessing processes. Section 4 presents experiment results with mixed-frequency forecasting models on export data. Section 5 closes with a discussion on the results and future works.

2   Related Works

The forecasting microeconomic indicators in general and export, in particular, are very interested and studied by many researchers. In which most studies use data with the same frequency with many different methods such as the gravity model, export demand model, Bayesian regression model, etc. and machine learning methods such as Neural networks, fuzzy set theory, genetic algorithms.

From the perspective of international trade, many export forecasting models approach from classical to modern theory. The classical theoretical models for export forecasting include the Ricardian model, the technology-based model, and the Heckscher-Ohlin resource factor model (Bussière et al. 2005). The endogenous growth and gravity models have also contributed to theoretical advances along with the theory of competitive advantage. In these models, factors such as foreign trade return to scale, monopolistic competition, diversity or market failure are considered to be included in the forecasting model (Feenstra 2015). Mehta and Mathur (2003) reviewed existing models for short-term forecasting of Indian exports in which export turnover depends on the import demand for domestically produced products of the partner countries, on the exchange rate and export price index and their lagged variables. Bussière et al. (2005) analysed the rapid trade integration of the Central and Eastern European countries (CEECs) with the euro area since 1980 and drew implications for further integration by using as a benchmark an enhanced gravity model. In this study, the results of the fixed-effects estimator and the dynamic OLS specification indicated that the economic size of neighbouring countries has a highly significant and less than the proportional impact on bilateral trade, the real exchange rate variables also enter the regression significantly. Similarly, having a common border and speaking the same language increase trade between the two countries, while the common territory dummy is not related to total trade. The authors also showed that the predicted trade values derived from the gravity model may be biased if not taking into account adjustments to standard trade conditions after the opening-up of Eastern Europe, which may translate into distorted estimates for the fixed effects.

Recently, according to research by Siggel (2006), one of the most commonly used models to forecast export turnover is the export demand model. This model assumes that supply is infinitely elastic, i.e. when demand exists, any supply can be produced. In the export demand model, most variables such as exchange rate, price index and the relative price of exports are used, in which relative price is one of the very important factors determining the competitiveness of export activities and comparative advantage. Stoevsky (2009) used the export demand model to forecast the total exports of Bulgaria, where the predictors in the model include the export price index of goods and services (nominal or real), the exchange rate (nominal or real), a composite index that measures external demand for domestically produced products, the price vectors of the economy's key exports, and their lagging variables. Lehmann (2015) forecasted the import-export turnover index of most European countries (including 18 Eastern and Central European countries, 28 EU countries) to study the economies of these countries. Survey-based variables (called soft variables) such as business confidence index, consumer confidence index, production expectation index, etc., are included in the import and export demand model to see if the forecast error is lower when compared with the model that includes only hard variables such as price variables and price indexes, aggregate demand variables, exchange rates, etc. However, the results in most countries show that adding a soft variable to the import and export forecasting model using the demand model method gives a much lower forecast. In fact, there are many other factors (or variables) that also affect the economy's exports, such as national production capacity, the skills of employees, inventories and credit balances of manufacturing industries, domestic and international political-economic situation, etc.

With the Bayesian regression model, Eckert et al. (2019) used a comprehensive data set containing Swiss exports to forecast exports. The all-time series covered the period from 1988 to 2018 with monthly frequency and was calculated in Swiss francs, they were not seasonally adjusted. In order to check the accuracy of all forecasts, the authors computed the log of the RMS forecasting error of the base forecasts relative to the MSE of the coherent forecasts from each method. The results showed values above zero indicate a better forecast performance. This also provided strong evidence that in addition to making coherent predictions, the Bayesian estimate helped to improve forecast accuracy.

In the approach using machine learning methods, Urrutia et al. (2019) forecasted Philippine imports and exports using Bayesian Artificial Neural Network and ARIMA model regression. In this study, data were collected from the Philippine Statistics Authority with a total of 100 observations in order to forecast values of imports and exports from the first quarter of the year 2018 to the fourth quarter of the year 2022 using the most fitted model. By computing and comparing the MSE, RMSE, NMSE, MAE, and MAPE of each model, the researcher deduced the fitted model that can use in forecasting the imports and exports of the Philippines is the Bayesian Artificial Neural Network. Upon using the Paired T-test, the p-value for the imports is 0.055 and for the exports is 0.054 which means that there is no significant difference between actual and predicted values for both imports and exports of the Philippines. This study could help the economy of the Philippines by considering the forecasted Imports and Exports which can be used in analyzing the economy's trade deficit.

Qu et al. (2019) forecasted import and export trends of Shandong province, China using the LSTM model (Long short term memory neural network). The study used monthly import and export data of Shandong province during the period from January 2001 to June 2018. Monthly import and export data often have large-scale, nonlinear and difficult to fit, so it is difficult to accurately forecast their trends. To solve this problem, the researchers proposed to use the LSTM model in combination with the cubic exponential smoothing method of the time series model. The results showed that the LSTM method gives a better result than the traditional time series regression model. Meanwhile, in order to forecast the demand of Chinese exports, Bin and Tianli (2020) based on artificial neural network theory, the theory of fuzzy systems and established forecasting models. Based on the system of export forecasting criteria determined on the basis of analyzing the research results of foreign trade export, the authors chose the evolutionary morphological neural network model to apply the export prediction and considered the influence of different factors, the obtained prediction results were ideal. On the other hand, Xie and Xie (2009) forecasted the total volume of import and export trade based on Gray forecast modelling (GM) which are optimized by a genetic algorithm. This paper introduced a combination prediction method based on fully utilize the advantages of the GM model and the characteristics of the genetic algorithm to find the global optimal combination, so the predictive model is more accurate. The results indicated that the model can be used as an effective tool to forecast the total volume of import and export trade.

For the above 3 studies, to forecast the total amount of a province's import-export trade, the authors performed on the data set with the same frequency, namely by month (in the study of Qu et al. (2019) or by year (as in the study of Xie and Xie (2009) and Bin and Tianli (2020).

Similarly, also with the same frequency of data by month, in 2020, Hai et al. built a model to forecast export turnover using the dimensional reduction method based on the Kernel trick. Different from traditional forecasting models, the study developed the model based on a large data set of 144 variables collected from various sources during the period from January 2013 to June 2019. By using a factor bridging equation model, factors were extracted from a large data set using a combination of attribute selection and attribute learning methods. The attribute selection method was used to remove the redundant and noisy information in the original data set while the attribute learning method was used in combination to extract the factors in the original data set after removing the noise or redundancy information. The test forecast results of the model built on the real data set of Vietnam showed that the out-of-sample prediction accuracy is about less than 3% compared with reality.

Trending of using different frequency data on the prediction of macroeconomic indicators.

The characteristics of macroeconomic indicators are that they are often published with different frequencies, some are published annually, but others are published quarterly, monthly, or daily. In addition to that, macroeconomic indicators often have different published times, making it difficult to exploit data for forecasting. To deal with those issues, one of the macroeconomic forecasting methods in recent years, which has been studied and applied by many scientists around the world, is models with variables collected and published with different frequencies. The research results show that this analytical method has the great advantage of optimally exploiting published data at different frequencies, possibly by day, month, quarter, or year. Some of the most popular models for analyzing data with mixed frequencies including the mixed data sampling (MIDAS) model, the bridge equations, and the mixed frequency VAR (MF-VAR) model. However, published studies have only focused mainly on applying mixed frequency data models for forecasting GDP, the indexes reflecting the growth of economies, and forecasting inflation without many forecasts for other important macro indicators such as exports. Meanwhile, studies in Vietnam have hardly mentioned the model of using datasets with mixed frequency in forecasting macroeconomic variables in general and exports in particular. Using mixed-frequency models for export forecasting is a large research gap. Therefore, this study investigates the performance of models that work on mixed frequencies data for the export forecasting problem.

3 Research Models and Data

3.1 Mixed-Frequency Models

Mixed frequency models are proposed with the aim of exploiting the availability of data at different frequencies and overcoming the disadvantages of the traditional VAR model when building a predictive model on a mixed frequency data set based on the VAR structure. One of the highest accuracy forecasting methods on datasets with different frequencies is the Kalman filter thanks to its ability to estimate the past, present, and future. However, it requires a lot of calculations, this method is not widely applied in the field of financial economics but instead in the field of automatic control. Besides, there are other popular mixed frequency methods such as the bridge equation model, the group of mixed data sampling model (MIDAS), and the mixed frequency VAR model (MF-VAR). These methods belong to the group of dynamic factor models and are applied in many fields of finance and economics (Kuzin et al. 2011; Foroni and

Marcellino 2013). In which, the bridge equation model and the MIDAS model are methods that approach single equations, i.e. have only one dependent variable, and MF-VAR is the model that approaches the system of equations in which all variables are dependent variables. Therefore, MF-VAR is often limited by the dimensional curse/ the curse of dimensionality, and determining the optimal delay of MF-VAR is often not as accurate as other methods. In this article, we study and apply two models MF-VAR and MIDAS to forecast Vietnam's exports based on collected data sets in the period from 2006 to 2020.

- *Mixed-frequency VAR model (MF-VAR)*

The MF-VAR model is a VAR model that analyzes time series datasets with mixed frequencies. This model is based on the state space approach and was proposed by Mariani and Murasawa in 2010 (Foroni and Marcellino 2013). In this model, to process data with different frequencies, the variables with low frequency is considered as a high-frequency variable with missing observations. Then, the low-frequency variables will be applied techniques to transform to the same frequency of the highest frequency variables in the model (for example, the Kalman filter is applied to estimates missing observations). MF-VAR usually has many parameters that are estimated by two following methods. The first method basing on the classical approach to estimate by using the maximum-likelihood technique while the second method is to use Bayesian inference with the expectation-maximization algorithm.

According to Mariano and Murasawa (2010), with the classical approach, when considering the VAR model in the state space, the disaggregation of the variable with the quarterly frequency y_{t_m} observed at time $t_m = 3, 6, 9, \ldots T^m$ into the unobserved monthly variable $y_{t_m}^*$ is described by the following aggregation equation:

$$y_{t_m} = \frac{1}{3}y_{t_m}^* + \frac{2}{3}y_{t_m-1}^* + y_{t_m-2}^* + \frac{2}{3}y_{t_m-3}^* + \frac{1}{3}y_{t_m-4}^*$$

This equation comes from the assumption that the series y_{t_m} is the average of the series $y_{t_m}^*$ with its lagged variables. Then, the process VAR(p) is represented as the following equation:

$$\phi(L_m)\begin{pmatrix} y_{t_m}^* & \mu_y^* \\ x_{t_m} & \mu_y \end{pmatrix} = u_{t_m}$$

Where

x_{t_m} and $y_{t_m}^*$ are variables with high frequency.
u_{t_m} is a random error with zero expectation and constant variance.

The parameters in the model are interpreted and estimated based on the state-space model with the maximum-likelihood estimation. According to Vladimir Kuzin et al. (2011), MF-VAR is more effective in long-term forecasting than nowcasting.

- *Mixed-data Sampling Model MIDAS*

MIDAS model is an analysis model with mixed frequency data, proposed by Eric Ghysels, Arthur Sinko & Rossen Valkanov in 2002, in which the explanatory variables have different frequencies that are equal or higher than the frequency of the dependent variable. In this model, to the explanatory variables with higher frequency, lagging polynomials are used to prevent an increase in the number of parameters as well as problems related to lag selection of the independent variables in the model.

The basic MIDAS model for a single explanatory variable and h_q- step- ahead forecasting, with $h_q = h_m/m$ is defined by:

$$y_{t_q + h_q} = y_{t_m + h_m} = \beta_0 + \beta_1 b(L_m; \theta) x_{t_m + w}^{(m)} + \varepsilon_{t_m + h_m}$$

In which:

y_{t_q} is a dependent variable with low frequency and x_{t_m} is an explanatory variable with high frequency.

t_q is the time index, $t_q = 1, 2, \ldots, T_q^y$, in which T_q^y is the final time that y_{t_q} have available data and $t_q + h_q$ is the time to be forecast.

t_m is the time index with higher frequency, $t_m = 1, 2, \ldots, T_m^x$, in which T_m^x is the final time that x_{t_m} have available data.

m is the index to defines a higher degree of frequency of the independent variable than the dependent variable. For example, if y has a quarterly frequency and x has a monthly frequency, then m = 3, and if y has a quarterly frequency and x has a weekly frequency, then m = 12.

w is the number of time periods based on high frequency for which the explanatory variable has data available earlier than the dependent variable.

$b(L_m; \theta) = \sum_{k=0}^{K} c(k; \theta) L_m^k$ is lag polynomials with L_m is the lag operator that is defined by:

$L_m^k x_{t_m}^{(m)} = x_{t_m - x}^{(m)} . x_{t_m + w}^{(m)}$ is skip-sampled from the high-frequency indicator x_{t_m}.

$c(k; \theta)$ are the parameters with lagged variables of the independent variable to be estimated, where k is the lag index.

β_0, β_1 are the coefficients of the regression model.

ε is the error of the model.

One of the main issues of the MIDAS method is finding a suitable parameterization for the lagging coefficients $c(k; \theta)$ to prevent the parameter increase in the model, avoid the situation that too many parameters cause difficulty for model estimation. This problem is due to x_{t_m} has a higher frequency than y_{t_q}. Several weight schemes have been developed to overcome this problem, the most popular of which is the Almon polynomial, also known as the "Exponential Almon Lag". Specifically, the Almon scheme is expressed as follows:

$$c(k; \theta) = \frac{exp(\theta_1 k + \ldots + \theta_Q k^Q)}{\sum_{k=1}^{K} exp(\theta_1 k + \ldots + \theta_Q k^Q)}$$

where Q is the number of parameters of θ, or $\theta = (\theta_1, \theta_2, \ldots, \theta_Q)$ are the parameters to be estimated. This function is quite flexible and can take various shapes with only a few parameters. Ghysel et al. (2004) applied this functional form with two parameters, that allows great flexibility and defines how many lags are included in the regression.

3.2   Research Data

- *Variable selection*

Based on international experience, the characteristics of export growth in Vietnam and the availability of the database, this research selects the variables for the model as shown in Tables 1, 2 and 3. For the low-frequency dependent variable, the research uses the quarterly export growth.

Variables representing the supply side include GDP growth (denoted by GDP_G), Net Foreign Direct Investment (NFDI), Industrial Production Growth Rate (IP_G); Variables representing the demand side include Retail sales growth (denoted by RS_G), Commodity import growth (denoted by CI_G), inflation rate (denoted by CPI); variables representing the export competitiveness include export commodity prices including exchange rate (current exchange rate and denoted by XR), weekly exchange rate (XR_W); The remaining variables represent the monetary sector, banking, stock market, futures contracts including Balance of payments (denoted by BOP), Capital account (denoted by KA), Money supply growth (denoted by MS_G), Reserve Ratio (denoted by RR), Composite Stock Price Index (denoted by CSPI), Interest Rate (denoted by IR), Total International Reserves (denoted by TIR), Balance of Trade (denoted by BOT). The inclusion of monetary variables also has important explanatory implications because Vietnam's import-export enterprises in general are highly dependent on bank capital. Large fluctuations in the stock market and futures market have a significant impact on export activities.

To forecast the quarterly export growth rate, the study uses a dataset of 22 variables with 19 economic indicators, of which: 5 quarterly frequency variables, 14 monthly frequency variables and 3 weekly frequency variables. They are described in detail in the following tables:

Table 1. The quarterly-frequency economics indicators

Variable	Unit	Indicator
EX_G	(%)	Export growth
GDP_G	(%)	GDP growth
BOP	(%)	Balance of payments
NFDI	(%)	Net foreign direct investment
KA	(%)	Capital account

Table 2. The monthly-frequency economics indicators

Variable	Unit	Economics indicator	Variable	Unit	Economics indicator
RS_G	(%)	Retail sales growth	**TIR**	(%)	Total international reserves (excluding gold)
IP_G	(%)	Industrial production growth	**CI_G**	(%)	Commodity import growth
MS_G	(%)	Money supply growth	**BOT**	(%)	Balance of trade
RR	(%)	Reserve ratio	**CPI**	(%)	Inflation rate
CSPI	Index	Composite stock price index	**XR_M**	USD/VND	Monthly exchange rate (average)
IR	(%)	Interest rate	**G_M**	USD/OUNCE	The future contract of gold
XR	Index	Exchange rate	**O_M**	USD/Barrel	The future contract of raw oil

Table 3. The weekly-frequency economics indicators

Variable	Unit	Economics indicator
XR_W	USD/VND	Weekly exchage rate
G_W	USD/OUNCE	The future contract of gold
O_W	USD/ Barrel	The future contract of raw oil

- *Data collection*

The research is carried out on a dataset of macroeconomic indicators, collected with different frequencies (quarterly, monthly, weekly) from the websites of the General Statistics Office, the IMF, the World Bank, ADB, Bloomberg. The data preprocessing and result representation is performed on Excel, while the analysis and forecasting with models are done on Eviews 11.

- *Data characteristics and preprocessing*

Due to the lag in data publication, there are differences in the number of observations obtained of the indicators in the last year of the research period, 3 data series are missing information (Money supply growth, Reserve ratio, total international reserves excluding gold).

Some variables do not have published data by month, by quarter but have published data at a higher frequency such as day, week, month. Therefore, in this study, high-frequency data were used to generate lower frequency data for those variables. The low-frequency variables are calculated as the average of higher frequency variables.

For example, Vietnam's export growth in the first quarter of 2020 will be taken as the average of Vietnam's export growth data for January, February, and March of 2020.

After collecting and pre-processing, the dataset was divided into 2 datasets of 2 different stages. In which, the data set in the period 2006 to 2018 is used to estimate the parameters in the regression models; the data set from 2019 to 2020 is used to forecasts.

- *Missing data issue*

In the 22 mentioned variables, 3 variables have missing data at the end of the research period, including Money supply growth, Reserve ratio and Total international reserves excluding gold. Typically, Money supply growth, Reserve ratio misses data of 12 months in 2020 and Total international reserves excluding gold misses data of 3 months of the fourth quarter of 2020. To overcome the issue of missing data, this research makes predictions for the missing values. Before forecasting for missing values and forecasting monthly export growth for Vietnam, the authors considered the correlation between independent variables of the same frequency in the three groups by the Pearson correlation coefficient and checked the stationarity of the data series by the unit root test method. The results showed that the data series are stationary, there are some series that are not stationary but overcome by taking the first-order difference.

- *Forecast for missing data (monthly frequency)*

The ARIMA model is one of the most effective short-term forecasting models for macroeconomic indicators. Therefore, this study applied the ARIMA (p, d, q) model to predict for 3 series with missing data. After considering the correlation schema of 3 series MS_G, RR, TIR, combined with the results of testing the stationarity of the series, the orders of AR (p), MA (q) and the order of difference (d) for stationary series have determined. Then the study regressed the different ARIMA models based on p, d, q levels and compared the models according to AIC and SIC standards to find the most suitable model. The results obtained are as follows (Table 4):

Table 4. The selection result of the ARIMA model

	MS_G	RR	TIR
P	2	1	1
D	1	1	1
Q	4	2	2
ARIMA (selected)	**(2, 1, 0)**	**(1, 1, 0)**	**(1, 1, 0)**
MAE	4.223	2.109	8.899
RMSE	5.835	2.577	11.183

(Source: The analysis result of authors)

4   Forecast Results

4.1   Overview of Real Quarterly Export Growth of Vietnam in the Period 2006–2020

From 2006 to now is the period of extensive economic integration with the important event that Vietnam became the 150th member of the WTO. As a result, in line with prior foreign trade policies and strategies, Vietnam's orientation is to continue leveraging its relative advantages to exploit present export markets while also developing new markets to support economic restructuring[1]. Shifting the export structure in favour of encouraging the export of high-value-added products, processed and manufactured goods, items with advanced technologies and brain content, while gradually lowering raw exports.

Figure 1 depicts the outcomes of Vietnam's quarterly export growth over this time period (Table 5).

Table 5. Descriptive statistics on quarterly export growth of Vietnam in the period 2006–2020

Observations	Mean	Median	Maximum	Minimum	Std. Dev
60	16.77	15.39	40.85	−19.6	12.08

(Source: data analysis results of the authors)

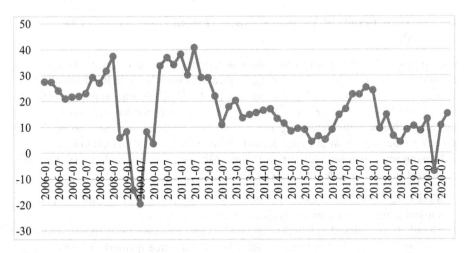

Fig. 1. Vietnam's quarterly export growth for the period 2006–2020 (Source: data analysis results of the authors)

---

[1] https://moit.gov.vn/tin-tuc/hoat-dong/xuat-nhap-khau-dong-luc-quan-trong-cho-tang-truong-kinh-te-d.html accessed on 01/08/2021.

Figure 1 shows that quarterly export growth has fluctuated throughout time. In particular, from 2008 to 2010, Vietnam's exports underwent significant fluctuations, dropping in the first quarter of 2008, hitting a low point in the third quarter of 2009 with a negative growth rate of almost −20%, and then beginning to recover; in the third quarter of 2011, Vietnam's export growth reached the highest level of about 41%. However, there were signals of a downturn starting in the fourth quarter of 2011, and this trend maintained from 2011 to the first half of 2016. Exports showed signs of increasing again from the end of 2016, reaching a new high in the fourth quarter of 2017 with a staggering 26.3% increase, but by 2018 they had reverted to the level of 2015–2016. Due to the impact of Covid-19 in the second quarter of 2020, Vietnam's export growth marked a serious decrease with a growth rate of −7%. However, the average export per quarter of the whole period 2006 to 2020 increased by around 16.8%.

When we examine the trend closely, we can see that the two periods with the lowest export value are the third quarter of 2009 (−19%) – the period when Vietnam's economy was severely impacted by the global financial crisis – and the second quarter of 2020 (−7%) – the period when Vietnam is facing the second wave of Covid-19, which means that shocks can have a significant impact on the country's exports. Furthermore, Vietnam's export growth trend is not stable, which further confirms the importance of forecasting. These actual results will be compared to the results obtained by the research team when conducting regression of mixed frequency data in Vietnam, which will be analyzed and presented in the following section.

4.2   Forecast Results of Vietnam's Export Growth Using the MF-VAR Model

The MF-VAR model is used in this study to predict Vietnam's export growth. There are some economic indicators at varying frequencies are included in the data collection. As a result, in order to develop the optimal model for predicting export growth, we examine the MF-VAR model using three scenarios in which the data of some economic indicators change with increasing frequency. This analysis also aims to see if the projected outcomes from utilizing the MF-VAR model with data containing multiple high-frequency variables are better or not. All three models include 16 variables mentioned in the research data, in which:

- Model 1: includes 5 quarterly frequency variables and 11 monthly frequency variables, with a lag after analysis and selection at level 2.
- Model 2: includes 2 quarterly frequency variables and 14 monthly frequency variables. The difference from model 1 is that the three quarterly frequency variables have been replaced by three monthly frequency variables including gold price, USD price and crude oil price.
- Model 3: includes 2 quarterly frequency variables, 11 monthly frequency variables and 3 weekly frequency variables. The difference between model 1 and model 2 is that model 3 includes datasets with all quarterly, monthly and weekly frequencies.

To compare the prediction results of the three models, the accuracy (or error) of the results must be measured. There is no one-size-fits-all indicator, therefore this is a difficult assignment. Currently, MAE (Mean absolute) and RMSE (Root mean squared) are regarded as two excellent predictive accuracy measurement criteria that are commonly applied in many studies. As a result, we examine the predicted accuracy of three models using those two criteria in this study (Table 6).

Table 6. Evaluation of prediction error of MF-VAR models

Time	Export growth (Quarterly)	MF_VAR1	Absolute error	MF_VAR2	Absolute error	MF_VAR3	Absolute error
2019Q1	4.34	24.17	19.83	20.56	16.22	43.28	38.94
2019Q2	9.15	24.69	15.54	20.67	11.52	47.62	38.47
2019Q3	10.52	25.23	14.71	20.76	10.24	50.71	40.19
2019Q4	8.70	25.72	17.02	20.81	12.11	40.47	31.77
2020Q1	13.21	23.14	9.92	20.82	7.61	37.88	24.66
2020Q2	−6.94	26.47	33.41	20.81	27.75	24.78	31.72
2020Q3	10.73	26.72	15.99	20.78	10.05	−2.71	13.44
2020Q4	15.22	26.93	11.70	20.76	5.53	−41.04	56.26
R2		**0.82**		**0.85**		**0.85**	
RMSE		**11.05**		**11.00**		**21.05**	
MAE		**7.20**		**7.79**		**16.88**	

(Source: data analysis results of the authors)

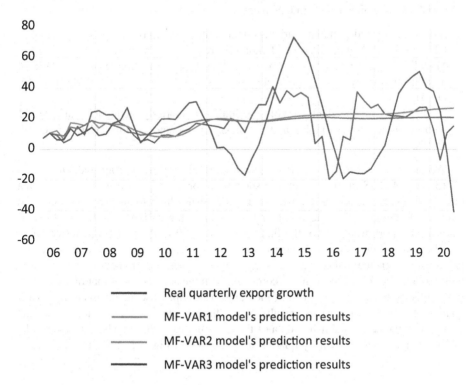

Fig. 2. Vietnam's real quarterly export growth and 3 forecast results *(Source: data analysis results of the authors)* (Color figure online)

Figure 2 shows that model 1 (orange line) delivers the closest predicted results to the true data series (blue line) across the whole period, with the smallest RMSE and MAE error values among the three models. Model 3 (red line) simulates the trend better, but the discrepancy between the model 3 forecast points and the original data is fairly large; there are variances of up to 30% in the fourth quarter of 2015, nearly 40% in Q4 2013, and more than 40% in Q4 2017.

The average error of the three models, according to MAE standards, is greater than 7%. Furthermore, because models 1 and 2 do not predict the peaks and troughs of the export growth cycle, determining the direction of recovery, growth, or decrease is difficult. Models 1 and 2's forecast values are almost ineffectual from 2012 onwards because export growth changes regularly with a significant margin, and model 3's amplitude fluctuates far beyond the actual value. As can be observed, the MF-VAR model is not suitable for forecasting Vietnam's export growth over the research period using the given collection of macroeconomic data.

4.3   Forecast Results of Vietnam's Export Growth by MIDAS Model

The MIDAS model has many different variations, this study applied the basic MIDAS model using the variables presented in Sect. 3.2. The research takes into account three variant models of MIDAS that are compared to the MF-VAR model, each consisting of 16 variables, of which 15 are independent.

- Model 1: 15 independent variables including 4 quarterly frequency variables: GDP_G, BOP, NFDI, KA and 11 monthly frequency variables.
- Model 2: 15 independent variables including 1 quarterly frequency of GDP; 14 monthly frequency variables (11 old variables and 3 new variables: XR_M, G_M, O_M).
- Model 3: 15 independent variables including 1 quarterly frequency variable, 11 monthly frequency variables and 3 weekly frequency variables (XR_W, G_W, O_W).

According to published researches, such as Kuzin (2011), Foroni and Marcellino (2013), the MIDAS model is useful in short-term forecasting since it is based on data that is closest to the forecast time. As a result, using high-frequency data collected through December 2020, the research team applies the MIDAS models to forecast Vietnam's export growth from the first quarter of 2019 to the fourth quarter of 2020.

The forecasting process is carried out as follows: Based on the data from the previous quarter, the author team will generate a forecast for the next quarter using a built model. The forecasted value is compared with the actual data of that quarter when it is published to estimate the error of the model. (Example: use the results of the fourth quarter of 2018 to forecast the results of the first quarter of 2019 and compare the results with the actual published data of the first quarter of 2019). The forecast results of Vietnam's export growth by quarter are presented in Fig. 3 and Table 7 below:

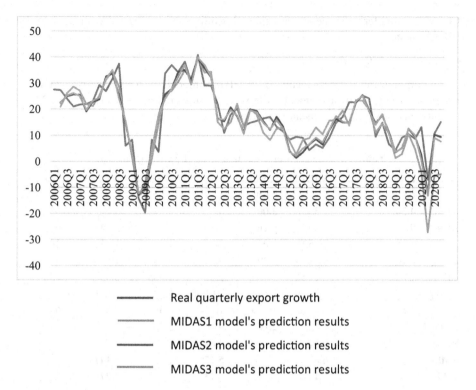

Fig. 3. Quarterly export growth forecast results of the model MIDAS1, MIDAS2, MIDAS3 *(Source: data analysis results of the authors)*

Figure 3 indicates that, when compared to the actual data series, the three MIDAS models accurately anticipate the variation trend of Vietnam's export growth chain. Most of the time, anticipated export growth follows real export growth, with only a few exceptions, such as the period early 2006 to mid-2007, the first two quarters of 2014, and 2015. The three forecast lines' range of fluctuations is fairly tiny in comparison to the value of the actual line, with the average absolute error of the three lines ranging from 4% to 5%. Furthermore, the forecasted outcomes closely track the export cycle's peaks and troughs. When comparing the prediction lines of the MIDAS2 and MIDAS3 models, it is clear that the two models provide very comparable forecast results; in fact, the two lines are practically identical.

The forecast results of the MIDAS2 and MIDAS3 models in the period 2019–2020 are pretty similar to the actual data when compared to model 1. Because of the advantage of using various and high-frequency data in the model, the mixed-frequency model outperforms classic models (VAR, SVAR, BVAR, VECM) in capturing rapid and unexpected changes in export activity. Traditional models, on the other hand, are frequently constrained by the use of data with the same frequency (for example, monthly models only use data with a monthly frequency, quarterly models only use data with a quarterly frequency), so their flexibility will be limited compared to the MIDAS model.

- *Evaluation of forecast errors*

To compare the forecast results of 3 MIDAS models, we also use two common predictive accuracy measurement criteria in many studies, namely MAE (Mean absolute) and RMSE (Root mean squared).

Table 7. Evaluation of prediction error of 3 MIDAS models

Time	Export growth (Quarterly)	MIDAS1	Absolute error	MIDAS2	Absolute error	MIDAS3	Absolute error
2019Q1	4.34	1.33	**3.00**	3.57	**0.77**	3.69	**0.65**
2019Q2	9.15	2.89	**6.26**	5.19	**3.96**	5.14	**4.01**
2019Q3	10.52	10.59	**0.06**	12.52	**1.99**	12.72	**2.20**
2019Q4	8.70	4.81	**3.89**	9.64	**0.94**	10.05	**1.35**
2020Q1	13.22	−4.056	17.27	2.57	10.65	0.65	12.56
2020Q2	−6.94	−27.09	20.15	−12.38	5.44	−13.08	6.14
2020Q3	10.73	−5.46	16.19	10.34	0.39	9.06	1.67
2020Q4	15.22	−4.84	20.06	9.46	5.76	7.69	7.53
R2		**0.82**		**0.82**		**0.81**	
RMSE		**6.87**		**5.17**		**5.141**	
MAE		**5.14**		**4.13**		**4.07**	

(Source: data analysis results of authors)

Table 7 shows that the three proposed MIDAS models with Almon weights provide a good forecast of Vietnam's quarterly export growth, particularly in the four quarters of 2019, the MAE of all three models are mostly smaller than the four quarters of 2020. In 2019, model 1 has 3 quarters, models 2 and 3 have all 4 quarters with forecast errors smaller than the average absolute error MAE, while the 4 quarters of 2020 mostly have errors bigger than MAE. If only considering model 1, the forecast for all quarters of 2019 is quite good with a low prediction error; the best forecast result belongs to the 3rd quarter with an error of only 0.06%, followed by the 1st quarter (3%) and the 4th quarter (3.89%); however, when entering 2020, the forecast error has a rather large difference, with all quarters having an error level of over 16%. For models 2 and 3, the scenario is similar, but the forecast error has improved significantly. If we only consider the year 2019, both models, in particular, have a low prediction error of roughly 4%; model 2 has up to three-quarters of anticipated results under 2%, while model 3 has three-quarters of forecast results under 2.2. Models 2 and 3 also have significantly smaller forecast errors than model 1 when calculated solely for 2020. This finding demonstrates that the MIDAS model is superior in short-term forecasting. The forecast error according to two criteria MAE and RMSE once again show that MIDAS2 and MIDAS3 models predict export growth of Vietnam better than MIDAS1 models, and the forecast results of two models (MIDAS2 and MIDAS3) is almost equally good.

This is due to the fact that the MIDAS2 and MIDAS3 models used more variables with higher frequency data than the original model. Specifically, three quarterly financial and monetary indices (capital account, balance of payments, and net direct investment) are replaced by three monthly monetary indices (gold price, USD/VND exchange rate, and crude oil price) in the MIDAS2 model, and three weekly monetary indices (gold price, USD/VND exchange rate, and crude oil price) in the MIDAS3 model. However, the sets of economic indicators included in the export growth forecast model are the same in models 2 and 3, with the exception of the monthly and weekly frequency, thus the difference in forecast results is not considerable, but the MIDAS3 model still outperforms the MIDAS2 model. At the same time, this forecast result demonstrates that the Midas model outperforms the MF-Var model in terms of predictive efficiency.

This result proves that the development of these forecasting models is on track, and the input data is suitably chosen. Furthermore, adding high-frequency financial data to the MIDAS regression model also enhanced prediction accuracy, demonstrating that financial data has an important role to play in predicting export growth. This finding implies that in order to fully utilize the forecasting ability of financial indicators, they must be combined with macroeconomic data.

5 Conclusion and Future Work

The study used the MF-VAR and MIDAS models to anticipate Vietnam's export growth from 2006 to 2020 using a data collection of varied frequency economic indicators. The analysis results show that the MF-VAR model has high forecasting error and is not suitable for forecasting the export growth of Vietnam, whereas the MIDAS model produces good forecast results using the same data set. Furthermore, as with several previously published study findings, the forecast results suggest that the MIDAS model is effective for short-term forecasting. The findings also suggest that high-frequency financial variables can be used to forecast Vietnam's export growth. This could be linked to the recent development of Vietnam's financial market.

From a policy standpoint, our findings imply that financial variables need to be closely monitored to predict fluctuations in the export cycle. In terms of modelling, our results point to the importance of linking financial sectors and economic realities in macroeconomic models. The role of financial variables in predicting export growth is not only due to their forward-looking nature but also to the close association between financial markets and a country's import and export activities.

In recent years, mixed frequency data analysis models have gotten a lot of attention and use in the economic and financial disciplines, particularly in forecasting national macroeconomic indicators. However, in Vietnam, there is almost no research in this field. Therefore, the application of models with mixed frequency data such as MF-VAR and MIDAS to forecast Vietnam's macroeconomic indicators is a promising new research direction, but due to the lag in publication of data on economic indicators, lead to gathering data for the study is still difficult, the research team has limited itself to estimating Vietnam's export growth at quarterly level, rather than at a higher frequency such as monthly or weekly.

Short-term forecasts of macroeconomic indicators will be critical in formulating policy and presenting each country's economic development strategies. As a result, with the goal of making timely and relevant forecasts in the future, the research team will continue to build and deploy models with varying frequency data to anticipate macroeconomic indicators.

Acknowledgments. This research is a part of the project funded by the Ministry of Education and Training and Thuongmai University under grant number B2020-TMA-04.

References

Başer, U., Bozoğlu, M., Eroğlu, N.A., Topuz, B.K.: Forecasting chestnut production and export of Turkey using ARIMA model. Turk. J. Forecast. **2**(2), 27–33 (2018)

Bin, J., Tianli, X.: Forecast of export demand based on artificial neural network and fuzzy system theory. J. Intell. Fuzzy Syst. (Preprint), 1–9 (2020)

Bui, T.M.N., Nguyen, T.Q.N., Nguyen, T.Q.C.: Using Arima model in forecasting the value of Vietnam's exports. J. Financ. Account. Res. **186**(1), 58 (2019)

Bussière, M., Fidrmuc, J., Schnatz, B.: Trade integration of Central and Eastern European countries:lLessons from a gravity model. ECB Working Article, No. 545 (2005)

Dung, T.T.M., Tuan, V.T., Huong, D.T.M., Luan, N.M.: Applying quantitative methods in forecasting Pangasius exports. Sci. J. Can Tho Univ., 123–132 (2014)

Eckert, F., Hyndman, R.J., Panagiotelis, A.: Forecasting Swiss exports using Bayesian forecast reconciliation (No. 457). KOF Working Papers (2019)

Feenstra, R.C.: Advanced International Trade: Theory and Evidence. Princeton University Press, Princeton (2015)

Foroni, C., Marcellino, M.G.: A survey of econometric methods for mixed-frequency data. Available at SSRN 2268912 (2013)

Ghysels, E., Santa-Clara, P., Valkanov, R.: The MIDAS touch: mixed data sampling regression models. Powered by the California Digital Library University of California (2004). 34p. https://escholarship.org/uc/item/9mf223rs

Hai, N.M., Dung, N.D., Thanh, D.V., Lam, P.N.: Building export value forecast model using dimensional reduction method based on kernel technique. In: The XXIII National Conference: Some Selected Issues of Information and Communication Technology (2020)

IMF: IMF Country Report No. 21/42 (2021)

Kuzin, V., Marcellino, M., Schumacher, C.: MIDAS vs. mixed-frequency VAR: Nowcasting GDP in the euro area. Int. J. Forecast. **27**(2), 529–542 (2011)

Le, N.B.N., T. A., Le, L.Q.T.: Forecast model for shrimp export prices of Vietnam. Sci. J. Can Tho Univ., 188–195 (2018)

Lehmann, R.: Survey-based indicators vs. hard data: what improves export forecasts in Europe? (No. 196). ifo Working Paper (2015)

Mariano, R., Murasawa, Y.: A coincident index, common factors, and monthly real GDP. Oxford Bull. Econ. Stat. **72**(1), 27–46 (2010)

Pyo, H.K., Oh, S.H.: A Short-term Export Forecasting Model using Input-Output Tables (2016)

Qu, Q., Li, Z., Tang, J., Wu, S., Wang, R.: A trend forecast of import and export trade total volume based on LSTM. In: IOP Conference Series: Materials Science and Engineering, vol. 646, no. 1, p. 012002. IOP Publishing (2019)

Schorfheide, F., Song, D.: Real-time forecasting with a mixed frequency VAR. Mimeo (2011)

Siggel, E.: International competitiveness and comparative advantage: a survey and a proposal for measurement. J. Ind. Compet. Trade **6**(2), 137–159 (2006)

Stoevsky, G.: Econometric Forecasting of Bulgaria's Export and Import Flows. Bulgarian National Bank Discussion Articles DP/77/2009 (2009)

Nguyen, T.N.A., Pham, T.H.H., Vallée, T.: Trade volatility in the association of Southeast Asian nations plus three: impacts and determinants. Asian Dev. Rev. **37**(2), 167–200 (2020)

Urrutia, J.D., Abdul, A.M., Atienza, J.B.E.: Forecasting Philippines imports and exports using Bayesian artificial neural network and autoregressive integrated moving average. In: AIP Conference Proceedings, vol. 2192, no. 1, p. 090015. AIP Publishing LLC (2019)

World Bank: World Development Indicator (2020)

Xie, Q., Xie, Y.: Forecast of the total volume of import-export trade based on grey modelling optimized by genetic algorithm. In: 2009 Third International Symposium on Intelligent Information Technology Application, vol. 1, pp. 545–547. IEEE (2009)

Improving Morphology and Recurrent Residual Refinement Network to Classify Hypertension in Retinal Vessel Image

Vo Thi Hong Tuyet[1,2,3] and Nguyen Thanh Binh[1,2(✉)]

[1] Department of Information Systems, Faculty of Computer Science
and Engineering, Ho Chi Minh City University of Technology (HCMUT),
268 Ly Thuong Kiet Street, District 10, Ho Chi Minh City, Vietnam
{vthtuyet.sdh19, ntbinh}@hcmut.edu.vn
[2] Vietnam National University Ho Chi Minh City,
Linh Trung Ward, Thu Duc City, Ho Chi Minh City, Vietnam
[3] Faculty of Information Technology, Ho Chi Minh City Open University,
Ho Chi Minh City, Vietnam
tuyet.vth@ou.edu.vn

Abstract. Prolonged or severe hypertension leads to vascular changes, resulting in endothelial damage and necrosis. The hypertension classification based on the retinal blood vessel segmentation is one of the state-of-the-art approaches. Therefore, the detection and classification of hypertensive retinal images is very useful in the diagnosis of hypertension. This paper proposed a method for the hypertension classification based on saliency, in which the extracting discriminative features keep information in the spatial domain. It includes three stages: morphology for enhancing the quality of the parameters, recurrent residual refinement network for feature extractions of the saliency and hypertension classification. The result of the proposed method is about 92.9% and better results than others in the STARE and DRIVE dataset.

Keywords: Morphology · Saliency · Recurrent residual refinement network · Hypertension classification · Segmentation

1 Introduction

Hypertension is a chronic disease where the pressure of the blood on the artery walls is increased. Hypertension puts a lot of pressure on the heart and is the cause of many serious cardiovascular complications, such as: cerebrovascular accident, heart failure, coronary heart disease, myocardial infarction, etc. The hypertension classification based on the retinal blood vessel segmentation is one of the state-of-the-art approaches. The input of the classification models is influenced on the main fields or parameters which include neighbor distance, gray level, boundary between areas together, the noise/blur values, etc. In medical image denoising, Nguyen [1] applied context awareness in the bandelet domain for ultrasound image. The nature of context awareness is linker and inference of around objects. Nageswara [2] develops an energy method of active contour for multi-level Laplacian filters (LP). The contourlet in image

© ICST Institute for Computer Sciences, Social Informatics and Telecommunications Engineering 2021
Published by Springer Nature Switzerland AG 2021. All Rights Reserved
P. Cong Vinh and N. Huu Nhan (Eds.): ICTCC 2021, LNICST 408, pp. 20–31, 2021.
https://doi.org/10.1007/978-3-030-92942-8_2

had used direct filter bank (DFB) for the reversing process. Some transforms of wavelet domain or spatial also combine with LP, DFB or multi-level of the decomposition. However, adapting with a wide range of conditions gives the lost information.

Morphology method overcomes this dissolution of transforms by shape, the relationship between neighbor pixels, etc. Gehad [3] proposed mathematical morphology to segment the retinal blood vessel. Zhiyong [6] used the Bayesian filter with the construction for curves to overcome the connection of edge maps for segmentation. The decomposition into domain and reversing by curvelet coefficients with morphological reconstruction had connected in fundus segmentation [4]. The clustering and operator about shape were also presented in [5, 8]. Elaheh [7] showed the morphology as a component analysis for detection. All the approaches had been the mathematical for curves or neighbor pixels of multi-level on the color channel. The limitation of these methods is not enough parameters of the object to use. As a result, the duplication or likeness creates a hard challenge for segmentation.

In recent years, researchers often use neural networks and deep learning for the feature object extraction. U-Net [9] and Fully Convolutional Network (FCN) [10] are the delegates of them. Mostafiz [11] used encoder and decoder of U-Net architecture with fuzzy inference system for retinal blood vessel segmentation. Besides that, hypertension risk diagnosis based on neural networks and fuzzy systems were also concerned in [12]. However, these models had influenced absolutely the parameters. If position moves or changes, the learning results will be affected. Deep multi-level networks for saliency classification [13–15] were proposed for end-to-end learning at weak objects. Recurrent Residual Refinement network (R3Net) with multiple-stage refinement in low-level and high-level features gives the deep supervision for initial classification.

In recent methods, adapting morphology or the learning model for strong area detection has been done. In [16], CNN in shearlet domain for segmentation was used in content-based medical image retrieval. The issue in classification was solved by morphology and power transmission line [17], CNN and morphology [18] and region growing [19]. Superpixels with clusters for saliency and EfficientNet were the idea of Mehmood [20] and Umit [21], respectively. The primary characteristic of connection is the multiple-morphological and superpixel to grow or to update the window size of classification. From the above algorithms, the combination of deep learning for saliency gives better segmentation and classification. However, the dividing low-level and high-level are not only refinement but also using supervision.

The abnormal or disease detection is necessary for the treatment. Hypertension is one of the pressing issues in diseases because of the other comorbidities. The dangers which are created from hypertension are terrible for patients. Therefore, predicting hypertension is necessary. The information for risk detection from retinal blood vessels is very enormous, and hypertension is also. The sensitivity of other areas, the opposition of retinal and blood are the vital parameters for the learning models. Patricia [12] applied fuzzy for diagnosis in neural networks. Yongbo [22] proposed a method to use morphological features of photoplethysmography which were extracted in relation between physiological and systolic blood pressure in depth. The retinal fundus photography is the dataset in [23] for predicting hypertension based on deep learning. Continued with this job, the evaluation of the outcomes of hypertension [24] based on physical indicators and predicted by cross-validation. Xue [25] proposed to use the Gaussian net model with saliency for retinal vessel segmentation.

From the above analysis, the benefit of physiology between an object's structure and morphology with deep learning for saliency classification in hypertension classification is doable. However, choosing the morphology model is still complicated. This paper proposed a method for hypertension classification by improving morphology and R3Net in retinal vessel image. The rest of the paper is organized as follows: Sect. 2 presented the background about saliency classification, the proposed method for classification presented in Sect. 3. Section 4 and 5 are the experimental results and conclusions, respectively.

2 Saliency Classification Background

Neural network for saliency classification [13] used a non-linear combination of features extraction which collected the final convolutional layer of deep multi-level network. From low and high weights level, the feature maps would be natural for classification. In some research, the architecture for saliency classification can be divided into multi-level encoder (convolution layer). The high-level receptive fields are the semantic information, such as: each part in object, color channel, gray level in surface, etc. The low-level features can be extracted as: edges, corners, orientation, etc. A sample system for saliency classification can be presented as Fig. 1.

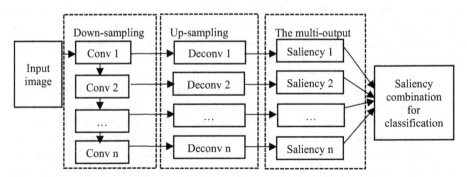

Fig. 1. A deep learning for saliency classification

Each layer has weight (w) and height (h). Deconvolution is a dimension with w × h with the input images. However, the size of the saliency map also gives the similarity. Loss function with cross entropy is calculated as Eq. (1):

$$H(p, q) = E_p - \log q \tag{1}$$

where, p and q are two distribution values. In some classification, if there are C classes for the dataset, the loss function will be defined as Eq. (2) with pixel x and vector of j element and W matrix.

$$J(W; x, y) = -\sum_{j=1}^{C} y_{ji} \log(a_{ji}) \tag{2}$$

3   Hypertension Classification Based on the Morphology and R3Net

This section clearly presents the proposed method for classifying hypertension. The proposed method presented as Fig. 2, includes the stages as: morphology, recurrent residual refinement network and saliency classification for hypertension classification.

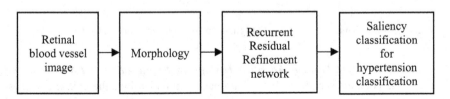

Fig. 2. The proposed method for hypertension classification

3.1   Morphology Stage

In the first stage in Fig. 2, the morphology is done with the green color of the retinal blood vessel. This stage is presented as Fig. 3. Its steps are explained below.

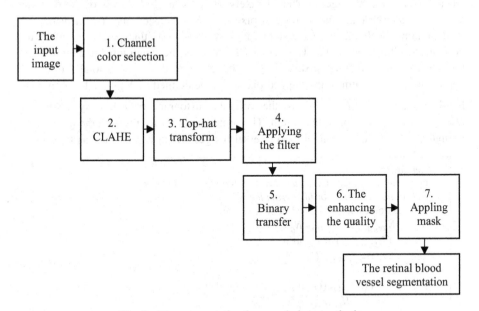

Fig. 3. The segmentation by morphology method.

+ *Firstly, Color Channel Selection.* In RGB (red, blue, green), the saturated is red, the influences by lighting, the high contrast between blood vessels and background are blue and green color, respectively. This step only selects the green channel as Fig. 4.

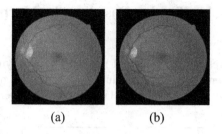

(a)   (b)

Fig. 4. Color channel selection. (a) Original image input, (b) Green channel selection.

+ Secondly, Contrast Limited Adaptive Histogram Equalization (CLAHE). In the simple term, a distinct section in the global image will redistribute the lightness values. Original intensity image divided into non-overlapping contextual areas. The area pixel for preserving is 8×8 pixels. These areas are converted into gray levels and accorded the histogram in each contextual. Then, the contrast limited histogram is calculated by Eq. (3):

$$N_{avg} = (NrX \times NrY)/N_{gray} \tag{3}$$

where, N_{avg} is the average number of pixels, N_{gray} is the gray levels of the dividing areas, NrX and NrY are the number of pixels for X-dimension and Y-dimension.

Continued with the normalized in range from 0 to 1, this period is called the normalized Clip Limit (CL). The actual CL, N_{CL}, is the condition for clipping histogram and the remaining pixels. N_{CL} is the product of N_{avg} and total number of clipped pixels. The sum of the clipped pixels can be defined as $N_{\sum clip}$. The condition for histogram with $H_{region}(i)$ is the original histogram at i-th gray level and $H_{region\_clip}(i)$ is the clipped histogram. The pseudocode for results with the input is the normalized CL value of retinal blood vessel images which is presented as follows.

Pseudocode:
Input: *the normalized CL value of retinal blood vessel images.*
Output: *the condition for histogram level*
    Function calculate_Hcondition(CL value)
        $N_{avggray} = N_{\sum clip} / N_{gray}$
        if $H_{region}(i) > N_{CL}$ then
            $H_{region\_clip}(i) = N_{CL}$
        else
            if $(H_{region}(i) + N_{avggray}) > N_{CL}$ then
                $H_{region\_clip}(i) = N_{CL}$
            else
                $H_{region\_clip}(i) = H_{region}(i) + N_{CL}$
            end if
        end if
        return H_{region}
    End function

The output of the above pseudocode is the condition for histogram level and keeps the strong information. Then, the probability density of each intensity value is calculated by Eq. (5):

$$p(y(i)) = \frac{(y(i) - y_{min})}{\alpha^2} \times exp\left(-\frac{(y(i) - y_{min})^2}{2\alpha^2}\right) \tag{5}$$

where, α is a scaling parameter of Rayleigh distribution = 0.04, y(i) is the Rayleigh forward transform, the lower bound of the pixel value is y_{min}.

The reduction for the linear contrast can be calculated by division between the input value (minimum value of transfer function) and maximum value of transfer function. As a result, the updated artifacts base on the new gray level within each contextual area.

+ *Thirdly, The Top-Hat Transform.* It includes the mask making and the subtraction calculating process of the previous step with the created mask. Its opening is dependent on the square structuring element, the morphological activities of erosion and dilation.

+ *Fourthly, Gaussian Filter for Giving the Blur Values.* The distribution is defined as the standard deviation (σ), the distances from the origin to horizontal and vertical axis are x and y, respectively. Each pixel depends on the neighbor pixel which is detected by distance G (x, y). It is calculated as Eq. (6).

$$G(x, y) = \frac{1}{2\pi\sigma^2}\left\{-\frac{x^2 + y^2}{2\sigma^2}\right\} \tag{6}$$

+ *Fifthly, The Binary Period by Mean-C Thresholding.* The sliding window which has M × N size is edited for calculating the limit threshold by convolution. The subtraction with C constance is doing.

+ *Sixthly, Enhancing the Image Quality.* The proposed method uses a mask for denoising by contour detection. The contour list of an image has been found and continued for the next step. If the acreage boundary is more than 150, the mask list will be checked. This mask is used for denoising. The results from CLAHE step (step 2) to the denoising step (step 6) in Fig. 3 are presented in Fig. 5.

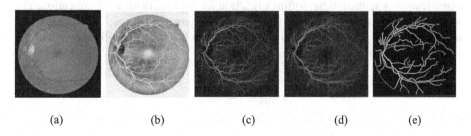

(a)   (b)   (c)   (d)   (e)

Fig. 5. The result of each step in the morphology method. (a) The result of the CLAHE step. (b) The result of the top-hat transform step, (c) The result of the Gaussian filter step, (d) The result of the Mean-C threshold step, (e) The result of the denoising step.

+ *Finally, Removing These Redundant Details Depends on a Binary Mask.* Any algorithm for the binary converting is also screened by the thresholding T and the gray level of the current pixels. In this step, the default value of T is 10. Then, the morphometric operator causes the highlights to narrow the dark regions. Figure 6 presented the final result of the segmentation based on the morphology method.

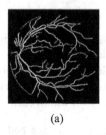   

(a)   (b)

Fig. 6. The result of the segmentation based on all steps of the morphology method. (a) The input of this process, (b) The final morphology method results.

3.2   R3Net and Saliency Classification for Hypertension Classification Stage

To classify hypertension retinal vessels, the proposed method only uses step 1, 2, 3, 4 and 5 in Fig. 3 (morphology method). So, Fig. 2 was modified and presented as Fig. 7.

The output of the first stage, R3Net has applied with multiple-stage refinement in the second stage. The architecture of R3Net for saliency map as follows in Fig. 7. Each block of conv has 3×3, with the sum of saliency map of each step with output of conv block 3 is 1×1 in residual results.

In the model, the morphology is divided into low-level and high-level with the current size. Each convolution block of residual refinement block (RRB) with two steps. The size of the input images is 512×512 pixels. If the size is more than the initial value, the overlap will be done. From the first period, the morphology with the Green color choosing which influence for the kernel size. The RRB in the first step gives the saliency map and that is the combination with the high-level matrix for the input of the first block in the next RRB step. ReLU in each conv block keeps the default size. The final output will be calculated with the summary. Then, saliency classification is applied to the softmax function for the hypertension classification. The loss function is calculated by Eq. (2). The dataset has two classes: hypertension and non-hypertension class. As a result, the final stage gives the classification about the kind of disease in retinal blood vessels.

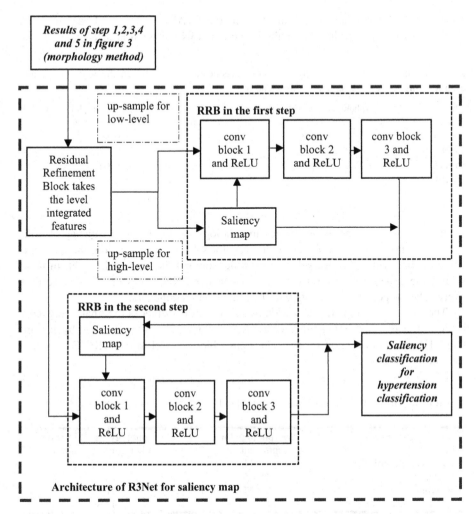

Fig. 7. Explained the Fig. 2

4 Experimental Results

Experimental programs were developed on a computer of Intel core i7, 3.2 GHz CPU and 16 GB DDR3 memory. The retinal blood vessels in the STARE [26] and DRIVE [27] dataset are used in this experiment. The number of images in the STARE and DRIVE dataset are 397 and 40 images, respectively. In the STARE dataset, the size of each image is 700 × 605 pixels for all disease images. This dataset divides 70% for training and 30% for testing. The size of each image in the DRIVE dataset with 565 × 584 pixels. Here, the experiment used 100% of this dataset for testing the hypertension classification.

Table 1 presented the evaluation values of the morphology method and some deep learning models: Convolutional neural network (CNN), Fully convolutional network

(FCN) method for saliency classification in the STARE and DRIVE dataset. The combination of the morphology method with our R3Net method is better than others.

Table 1. The evaluation of the hypertension classification (%) between the morphology and deep learning models for saliency classification.

Dataset/ The method	The morphology method	The morphology + CNN method	The morphology + FCN method	The morphology + R3Net method
STARE dataset	79.85	88.35	90.06	91.59
DRIVE dataset	82.71	89.17	91.05	92.90

In Table 1, the experiment applies morphology for the input of saliency classification, the morphology combined with CNN for saliency classification, the morphology combined with FCN for saliency classification and the proposed model. In these results, the experiment applies 6 blocks of encoder CNN, 6 blocks of FCN, respectively. The proposed method fits with hypertension classification.

The experiment implements a wide range of threshold/filter in the final step (step 7 in Fig. 3) of the morphology period and R3Net for saliency classification by Jaccard Index [28] in segmentation results. The retinal blood vessel in the DRIVE dataset is used in this testing.

Table 2. The average Jaccard index in threshold or filter step with the proposed model.

The method / Filter	Morphology for segmentation	Morphology and R3Net for saliency classification
Bayesian thresholding	79.34	90.72
Median filter	78.05	90.21
Gaussian filter	84.77	91.80
Mean-C thresholding	**86.24**	**93.50**

The Table 2 presented the average Jaccard Index in threshold or filter step with the proposed model. In Table 2, the second column (morphology for segmentation), the experiment applies the algorithm which has 7 steps in Fig. 3 for segmentation to evaluate clearly with saliency for classification (morphology without R3Net for saliency classification). The third column is the Morphology method combined with R3Net in testing. In two approaches, the Jaccard Index based on Mean-C thresholding for segmentation is higher than Bayesian thresholding, median filter, and Gaussian filter.

Table 3. The classification evaluation between the proposed method with the recent methods.

| Method

The accuracy (%)	R3Net for saliency classification [15]	Morphology-Based Feature-Extraction [17]	The morphological and superpixel based fast fuzzy C-mean clustering network for saliency [20]	The proposed method
The accuracy (%) of classification in STARE dataset	89.73	88.07	90.91	**91.59**
The accuracy (%) of classification in DRIVE dataset	90.10	90.64	91.97	**92.90**

Mean-C thresholding was applied for evaluation with R3Net for saliency [15], morphology-based feature-extraction [17], the morphological and superpixel based on fast fuzzy C-mean clustering network for saliency [20]. The final comparison is presented in Table 3.

From these values, using morphology for input of deep learning gives a better result for classification systems. However, depending on the filter or threshold for choosing the number of the input parameters is also a vital step. The results of the proposed method are better than the others. Because R3Net, which changes the parameters (such as color channel or window size of the first layer in blocks) makes the system improve the accuracy.

5   Conclusions and Future Works

Acute hypertension often causes reversible vasospasm in the retina, and malignant hypertension can cause papilledema. Prolonged or severe hypertension leads to vascular changes, resulting in endothelial damage and necrosis. Therefore, the detection and classification of hypertensive retinal images is very useful in the diagnosis of hypertension. This paper proposed a method to improve the morphology and R3Net for hypertensive classification in retinal vessel image. The nature of improving the quality and choosing the color channel for transform overcomes the limitation of the noise details in the learning models. The morphology is an old approach in image segmentation. However, if it becomes one of parameters of saliency classification and R3Net, the results will be better. The results of the proposed method compare with the results of R3Net for saliency classification method [15], Morphology-Based Feature-Extraction method [17], the morphological and superpixel based fast fuzzy C-mean clustering network for saliency method [20]. Experiment results of the proposed method are better than others in the STARE and DRIVE dataset. In the future, the

R3Net architecture will continue to improve and experiment on other datasets to crease the accuracy of the proposed method.

Acknowledgement. This research is funded by Vietnam National University Ho Chi Minh City (VNU-HCM) under grant number B2019-20-05. We acknowledge the support of time and facilities from Ho Chi Minh City University of Technology (HCMUT), VNU-HCM for this study.

References

1. Binh, N.T., Tuyet, V.T.H., Vinh, P.C.: Ultrasound images denoising based context awareness in bandelet domain. In: Vinh, P.C., Alagar, V., Vassev, E., Khare, A. (eds.) ICCASA 2013. LNICSSITE, vol. 128, pp. 115–124. Springer, Cham (2014). https://doi.org/10.1007/978-3-319-05939-6_12
2. Nageswara Reddy, P., Mohan Rao, C.P.V.N.J., Satyanarayana, C.: Brain MR image segmentation by modified active contours and contourlet transform. ICTACT J. Image Video Process. **8**(2), 1645–1650 (2017)
3. Hassan, G., El-Bendary, N., Hassanien, A.E., Fahmy, A., Snasel, V.: Retinal blood vessel segmentation approach based on mathematical morphology. Proc. Comput. Sci. **65**, 612–622 (2015)
4. Quinn, E.A.E., Krishnan, K.G.: Retinal blood vessel segmentation using curvelet transform and morphological reconstruction. In: 2013 IEEE International Conference on Emerging Trends in Computing, Communication and Nanotechnology, pp. 570–575. IEEE (2013)
5. Mehrotra, A., Tripathi, S., Singh, K.K., Khandelwal, P.: Blood vessel extraction for retinal images using morphological operator and KCN clustering. In: 2014 IEEE International Advance Computing Conference, pp. 1142–1146. IEEE (2014)
6. Xiao, Z., Adel, M., Bourennane, S.: Bayesian method with spatial constraint for retinal vessel segmentation. Comput. Math. Methods Med. **2013**, 1–9 (2013)
7. Imani, E., Javidi, M., Pourreza, H.R.: Improvement of retinal blood vessel detection using morphological component analysis. Comput. Methods Programs Biomed. **118**(3), 263–279 (2015)
8. Dash, J., Bhoi, N.: Retinal blood vessel extraction using morphological operators and Kirsch's template. In: Jiacun Wang, G., Reddy, R.M., Kamakshi Prasad, V., Sivakumar Reddy, V. (eds.) Soft Computing and Signal Processing. AISC, vol. 900, pp. 603–611. Springer, Singapore (2019). https://doi.org/10.1007/978-981-13-3600-3_57
9. Ronneberger, O., Fischer, P., Brox, T.: U-Net: convolutional networks for biomedical image segmentation. In: Navab, N., Hornegger, J., Wells, W.M., Frangi, A.F. (eds.) MICCAI 2015. LNCS, vol. 9351, pp. 234–241. Springer, Cham (2015). https://doi.org/10.1007/978-3-319-24574-4_28
10. Jiang, Z., Zhang, H., Wang, Y., Ko, S.-B.: Retinal blood vessel segmentation using fully convolutional network with transfer learning. Comput. Med. Imaging Graph. **68**, 1–15 (2018)
11. Mostafiz, T., Jarin, I., Fattah, S.A., Shahnaz, C.: Retinal blood vessel segmentation using residual block incorporated U-Net architecture and fuzzy inference system. In: 2018 IEEE International WIE Conference on Electrical and Computer Engineering, pp. 106–109 (2018)
12. Melin, P., Miramontes, I., Prado-Arechiga, G.: A hydrid model based on modular neural networks and fuzzy systems for classification of blood pressure and hypertension risk diagnosis. Exp. Syst. Appl. **106**(1), 146–164 (2018)

13. Cornia, M., Baraldi, L., Serra, G., Cucchiara, R.: A deep multi-level network for saliency prediction. In: 23rd International Conference on Pattern Recognition (2017)
14. Lai, B., Gong, X.: Saliency Guided end-to-end learning for weakly supervised object detection. Computer vision and pattern recognition (2017)
15. Deng, Z., et al.: R3Net: recurrent residual refinement network for saliency detection. In: Proceedings of the 27th International Joint Conference on Artificial Intelligence, p. 684–690 (2018)
16. Tuyet, V.T.H., Hien, N.M., Quoc, P.B., Son, N.T., Binh, N.T.: Adaptive content-based medical image retrieval based on local features extraction in shearlet domain. EAI Endorsed Trans. Context-aware Syst. Appl. **6**(17), e3 (2019)
17. Godse, R., Bhat, S.: Mathematical morphology-based feature-extraction technique for detection and classification of faults on power transmission line. IEEE Access **8**, 38459–38471 (2020)
18. Pasupa, K., Vatathanavaro, S., Tungjitnob, S.: Convolutional neural networks based focal loss for class imbalance problem: a case study of canine red blood cells morphology classification. J. Ambient Intell. Human. Comput. 1–17 (2020)
19. Ma, Z., et al.: Individual tree crown segmentation of a larch plantation using airborne laser scanning data based on region growing and canopy morphology features. Remote Sens. **12**(7), 1078 (2020)
20. Nawaz, M., Yan, H.: Saliency detection via multiple-morphological and superpixel based fast fuzzy C-mean clustering network. Exp. Syst. Appl. **161**(15), 113654 (2020)
21. Atila, U., Ucar, M., Akyol, K., Ucar, E.: Plant leaf disease classification using EfficientNet deep learning model. Ecol. Inf. **61**, 101182 (2021)
22. Liang, Y., Chen, Z., Ward, R., Elgendi, M.: Hyptertension assessment using photoplethysmography: a risk stratification approach. J. Clinic. Med. **8**(1), 12 (2019)
23. Zhang, L., et al.: Prediction of hypertension, hyperglycemia and dyslipidemia from retinal fundus photographs via deep learning: a cross-sectional study of chronic diseases in central China. PLoS ONE **15**(5), e0233166 (2020)
24. Chang, W., et al.: A machine-learning-based prediction method for hypertension outcomes based on medical data. Diagnostics **9**(4), 178 (2019)
25. Xue, L.-Y., Lin, J.-W., Cao, X.-R., Zheng, S.-H., Yu, L.: A saliency and Gaussian net model for retinal vessel segmentation. Front. Inf. Technol. Electr. Eng. **20**(8), 1075–1086 (2019). https://doi.org/10.1631/FITEE.1700404
26. STARE dataset: https://cecas.clemson.edu/~ahoover/stare/probing/index.html. Accessed 8 Jan 2021
27. DRIVE dataset: https://computervisiononline.com/dataset/1105138662. Accessed 8 Jan 2021
28. Tuyet, V.T.H., Binh, N.T.: Improving retinal vessels segmentation via deep learning in salient region. Springer Nature Computer Science Journal **1**(5), 1–8 (2020). https://doi.org/10.1007/s42979-020-00267-z

The Impacts of the Contextual Substitutions in Vietnamese Micro-text Augmentation

Huu-Thanh Duong[1(✉)] and Trung-Kiet Tran[2]

[1] Faculty of Information Technology, Ho Chi Minh City Open University,
Ho Chi Minh City, Vietnam
`thanh.dh@ou.edu.vn`
[2] Department of Fundamental Studies, Ho Chi Minh City Open University,
Ho Chi Minh City, Vietnam
`kiet.tt@ou.edu.vn`

Abstract. The deep learning models rely on a huge amount of annotated training data to learn multiple layers of the features or representations and also avoid overfitting. However, the annotated dataset is unavailable, especially for the low resource languages. Building them is a tedious, time-consuming and expensive task. Thus, data augmentation has been mentioned as a perfect approach to generate the annotated data from the limited data without user intervention. In this paper, we evaluate the importances and the impacts of the contextual words to enhance the training data based on a pre-trained model which we build based on the reviews extracting the e-commerce websites in Vietnamese. We experiment on the sentiment analysis problem to evaluate the effectiveness of our approach.

Keywords: Data augmentation · Deep learning · Sentiment analysis · Contextual substitution

1 Introduction

Data augmentation (DA) techniques first have been used in image classification to enrich the image training data such as rotate, scale, translate, noise, image mixup, etc. They are gradually equipped in natural language processing (NLP) for the semi-supervised learning approach and deep learning models with the limited training data, especially suitable for the low-resource languages as Vietnamese in order with overfitting avoidance.

Deep learning (DL) has emerged as a spotlight, has been used widely and achieved state-of-the-art performance in computer vision and NLP by learning multiple layers of features or representations of the data. Actually, deep learning has been developed from artificial neural networks appearing since the 1990s.

© ICST Institute for Computer Sciences, Social Informatics and Telecommunications Engineering 2021
Published by Springer Nature Switzerland AG 2021. All Rights Reserved
P. Cong Vinh and N. Huu Nhan (Eds.): ICTCC 2021, LNICST 408, pp. 32–39, 2021.
https://doi.org/10.1007/978-3-030-92942-8_3

In that period, the neural network training has only one or two layers (shallow neural networks) as if having more layers, the training is more complicated and expensive in computation. Based on Zhang et al. [1], the rapid growth of deep learning (deep neural networks) in recent years is as: (1) the availability of computing power due to the advances in hardware, (2) the availability of hug amounts of training data and (3) the power and flexibility of learning intermediate representations.

Therefore, one of the barriers of DL models relies on a large amount of the annotated training data, but these datasets are scarce, unavailable in a domain and collecting the annotated training dataset is time-consuming and expensive. DA is an effective solution to generate new training data being unchanged the labels and improves generalization of the DL models.

2   Related Works

There are many approaches proposed to augment training data such as lexicon substitution, sentence shuffling, generative text, back translation, syntax-tree transformation, word or sentence embeddings mixup.

Lexicon substitution techniques are the simple approaches such as shuffling the words or sentences of the original text, or swapping the sentences between the texts in the same label. Wei J et al. [3] proposed a simple approach to generate new data, including random replacement, random deletion, random insertion and random swap. Random replacement takes a random word from the original text and replaces it by a synonym based on the thesaurus, Random deletion removes a random word in the text with a probability p, Random Insertion takes a random word not being a stop word and gets one of its synonym words to insert into a random position in the text, Random Swap swaps randomly two words in the text. These techniques are simple, easy to implement, no need for any other resources and actually improve the accuracy of the algorithms in practice although they might generate meaningless sentences.

Back translation leverages translation machines to obtain new training data R. Sennrich et al. [9], A. Sugiyama et al. [10], M. Fadaee et al. [11]. Firstly, the text is translated from its original language into the intermediate language, then it will be back-translated into the original language to obtain new data which retains the meaning of the text. The translated text has never been exactly the same as the original text. This is easy to understand and implement, still retains the meaning of the original text, and achieves good performance. In [12] proves that this is an effective approach to generate new training data and improve the performance of the classifiers. The weakest point of this method is to need an effective translator.

Inspired by cropping and rotating techniques of data augmentation in image processing, syntax-tree transformation generates the dependency tree of the original text, then using some rules transforms it into the augmented text which retains the meaning, such as changing from active voice to passive voice. G. G. Sahin [13] used a dependency tree to remove dependency links and move the tree fragments around the root. In [12], the authors also boosted the performances of the classifiers.

Word Embedding is a crucial tool for machine learning and DL algorithms to convert text into vectors. The words which are the same context will be near each other in vector space. Ideally, vectors of words have the same meaning being near each other. There are two approaches for word embedding as frequency-based embeddings and prediction based embeddings, we can use prediction-based embeddings to find and replace a word by its similar word. Word embedding-based substitution technique also replaces a word by other contextual words by using the pre-trained word embeddings such as Word2Vec, GloVe, Bert, etc. to select the most similar words in the embedding space. This approach reduces the cost to build the annotated data, thesaurus, wordnets, ontology, etc. However, being a black-box approach, some opposite words are near each other in vector space such as "tốt" (good) and "tệ" (bad), so it may affect the performances of some problems such as sentiment analysis.

K. Liu [4] based on the semantic similarity in the context in Word2Vec model to train synonym substitution list and replace the unknown words in the original words with synonyms in financial public opinion. S. Kobayashi [5] proposed the contextual data augmentation for the labeled sentences, this offers a list of substitute words predicted by a bidirectional language model according to the context. In the experiment, the author proved this solution improved the classifiers using CNN (Convolutional Neural Network) and RNN (Recurrent Neural Network). Furthermore, in embedding space can apply the mixup methods. The mixup technique is the latest approach by combining the word or sentence embeddings in the flow of the text classification. L. Sun et al. [7] and O. Kashefi et al. [8] proposed the mixup approaches and proved it significantly improved the performance of deep learning models.

In this paper, we approach the contextual substitutions to obtain new data. The rapid growth of web technologies has created a huge amount of data on the internet, mining them helps to resolve many problems effectively in machine learning, especially for low-resource languages. Our main contribution builds a pre-trained model for the Vietnamese reviews by collecting reviews from e-commerce websites and performs various experiments to evaluate the impacts of the contextual substitutions via deep learning models.

The rest of this paper is organized as follows: Sect. 3 presents the proposed approach, Sect. 4 shows the experimental results, the conclusions and future works are presented in Sect. 5.

3 The Approach

Word2Vec (W2V), which is one of the first models of Word Embedding, is a method to represent a vector of a word based on around words (contextual words). This uses a neural network having two layers (input and output) and one hidden layer. There are two models: Skip-gram and CBOW (Continuous Bag of Word).

The input of the skip-gram approach is the current word and the output is a word which has the most relationship to the current word. The vectors of input and output layers are similar to the number of dimensions and have a form of one-hot vectors. The number of dimensions of the hidden layer is equal to the predefined size (embedding size) and less than the size of input and output layers. Output layer uses a softmax activation function for every part of the vector. CBOW is the same as Skip-gram, but input is a list of words (contextual words) of the current word and output is the most related word of the current word.

We have built a pre-trained model in Vietnamese with W2V based on the data as reviews extracted from e-commerce websites, this is used to determine the contextual words of the replacing words in the text. The first approach replaces n words in the text randomly with the contextual words for new samples.

Furthermore, we use $tf \times idf$ scores to choose the replacing words instead of random choices. The words which have low $tf \times idf$ scores are uninformative, so replacing them usually does not affect the meaning of the original text. The original text is tokenized and calculated the $tf \times idf$ scores based on the list of reviews (comments) extracted from the well-known e-commerce websites in Vietnamese such as the gioididong.com, tiki.vn, shopee.vn, etc. In order to obtain new training data, we conduct n times the replacement of the words ordered by $tf \times idf$ scores in ascending order.

Data augmentation with mixup has first been proposed by Lei Zhang et al. [1] for image classification and shown an effective approach. They generate the synthetic samples through linearly interpolating a pair of input images and their corresponding labels randomly as the following equations, where x_i, x_j are the input vectors, y_i, y_j are the corresponding labels and the λ is the mixup-ratio having the value between $[0, 1]$.

$$\hat{x} = \lambda x_i + (1 - \lambda)x_j \tag{1}$$

$$\hat{y} = \lambda y_i + (1 - \lambda)y_j \tag{2}$$

This approach is also equipped in NLP and achieves promising results (Hongyu Guo [6], L. Sun et al. [7]). The mixup is a cross-language approach based on embedding space, thus this is a potential approach for our problems. Our work injects the mixup layer to obtain the synthetic samples before feeding to the output layer to predict the results.

4   The Experiments

We use the datasets mentioned in [12] (dataset 1 (9280 positive, 6870 negative), dataset 2 (2380 strong positive, 2440 positive, 2380 negative), dataset 3 (15000 for each positive and negative), dataset 4 (5000 for each positive and negative)) and has gotten 100 samples of each polarity for training. Firstly, it conducts the preprocessing techniques into the datasets to remove some noise information, normalize and clean the text such as lowercasing, emojis substitution, negation handling, intensification handling. We have also collected above 100 thousands of reviews from the e-commerce websites, this dataset is used to build the pre-trained model as Word2Vec and calculate the $tf \times idf$ weight of a word to determine for replacement.

The next experiment conducts n times for each sample to generate n new samples, each time chooses m words randomly and is replaced by their contextual words. This method generates 1000 new samples from 100 original samples of each polarity. In order to evaluate the impacts of the replacing words, we apply $tf \times idf$ weight to choose the replacing words in a sample. The $tf \times idf$-based substitution only replaces the uninformative words with the low $tf \times idf$ scores by the contextual word in W2V. The final experiments, these methods incorporate a mixup embedding approach in the mixup layers above the output layer.

The deep neural network includes two hidden layers (with 16 hidden units each) and one output layer, the hidden and output layers use the `relu` and `sigmoid` activation functions respectively. Since the output is a probability the loss function is `binary_crossentropy`. The training is performed for 20 epoches in mini-batches of 512 samples. Table 1 is the accuracies of our approaches with standard deviation of 10 runs, where the first column performs with the original data (the baseline result), the second column replaces randomly with the contextual words, $tfidf$-based substitution is presented in the third column.

Table 1. The mean of the accuracies with standard deviation in 10 runs: (1) The original data (baseline results), (2) Random substitution by the contextual words, (3) $tfidf$-based substitution by the contextual words, (4) Random substitution by the contextual words + mixup, (5) $tfidf$-based substitution by the contextual words + mixup.

Datasets	(1)	(2)	(3)	(4)	(5)
Dataset 1	0.7636 ± 0.050	0.8443 ± 0.004	0.8396 ± 0.012	0.8482 ± 0.0012	**0.8530 ± 0.0008**
Dataset 2	0.4148 ± 0.056	**0.5155 ± 0.011**	0.4492 ± 0.047	0.4031 ± 0.0020	0.4262 ± 0.0001
Dataset 3	0.7418 ± 0.054	0.7798 ± 0.003	0.7821 ± 0.005	**0.8066 ± 0.0002**	0.7983 ± 0.0003
Dataset 4	0.7567 ± 0.020	0.7817 ± 0.006	0.7861 ± 0.007	**0.7911 ± 0.0002**	0.7849 ± 0.0006

Figure 1 shows the visualization of the results, where the x axis is the experimental datasets, the y axis is the mean of the accuracies over 10 runs, each dataset runs all our experiments (see the caption of Fig. 1). The accuracy of the approaches overcomes the baseline result (the first columns of each dataset). The $tf \times idf$-based substitution is greater than random-based substitution (dataset 3 and dataset 4) so selecting good words to replace the contextual words is an essential point and affects the quality of the augmented data. However, other cases are in the opposite (dataset 1 and dataset 2). As earlier mentioned, word embeddings is a black-box approach, the opposite words may be near each other in embedding space such as "hài_lòng" (pleased) and "thất_vọng" (disappointed). In social reviews, some important words are low $tf \times idf$ score,

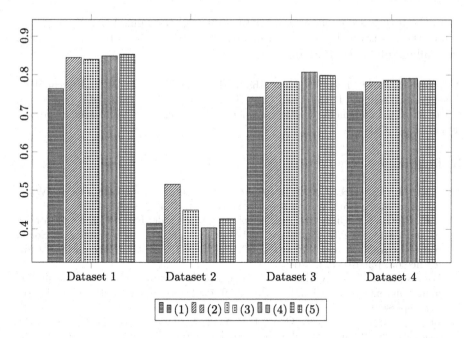

Fig. 1. The mean of the accuracies

for example the "hài_lòng" (pleased) word, which is commonly used in positive reviews, has low $tf \times idf$ score, so contextual words sometimes change the meaning of the text.

The accuracy of the contextual substitution which combines with the mixup embedding improves the performance of the datasets compared with only performing the contextual substitution. Although the accuracies decrease a little bit for dataset 2 (for two combinations) and dataset 4 (for combining with $tf \times idf$-based approach), the results of this combination are more stable with very minimum deviation for all datasets.

5   Conclusions and Future Works

This paper collects a set of reviews from well-known e-commerce websites and builds a pre-trained model for the contextual words. We conduct the various experiments to evaluate the effectiveness of the contextual substitution in generating new samples in Vietnamese. The experimental results show that selecting the word in the original texts and selecting a word in the contextual words for substitution affects the quality of the augmented data and the performances of the algorithms. The experiments have shown promising results, this utilizes the abundance of the internet data and can apply for cross languages, and saves cost to build the thesaurus or wordnets of the specific languages for the substitution approach.

In the future, we enhance more data to boost better quality and generalization of the pre-trained models and investigate novel methods to augment data, especially applying to Vietnamese.

Acknowledgment. I would like to thank the Center for Science and Technology Development Young - Thanh Doan City. HCM City (TST). This work is part of the "Vietnamese data augmentation for sentiment analysis based on deep learning" project supported by TST.

References

1. Zhang, L., Wang, S., Liu, B.: Deep learning for sentiment analysis: a survey. Wiley Interdisc. Rev.: Data Mining Knowl. Discovery **8**(4), e1253 (2018)
2. Han, S., Gao, J., Ciravegna, F.: Neural language model based training data augmentation for weakly supervised early rumor detection. In: Advances in Social Networks Analysis and Mining (ASONAM) 2019 IEEE/ACM International Conference on, pp. 105–112 (2019)
3. Wei, J., Zou, K.: EDA: Easy data augmentation techniques for boosting performance on text classification. In: ICLR 2019–7th International Conference on Learning Representations (2019)
4. Liu, K., Ergu, D., Cai, Y., Gong, B., Sheng, J.: A new approach to process the unknown words in financial public opinion. Proc. Comput. Sci. **162**, 523–531 (2019). https://doi.org/10.1016/j.procs.2019.12.019
5. Kobayashi, S.: Contextual augmentation: data augmentation by words with paradigmatic relations. In: Proceedings of the 2018 Conference of the North American Chapter of the Association for Computational Linguistics: Human Language Technologies, Vol. 2 (Short Papers), New Orleans, Louisiana, pp. 452–457 (2018). https://doi.org/10.18653/v1/N18-2072
6. Guo, H., Mao, Y., Zhang, R.: Augmenting data with mixup for sentence classification: an empirical study. ArXiv, arXiv:1905.08941 (2019)
7. Sun, L., Xia, C., Yin, W., Liang, T., Yu, P., He, L.: Mixup-transformer: dynamic data augmentation for NLP Tasks. In: Proceedings of the 28th International Conference on Computational Linguistics, Barcelona, Spain (Online), pp. 3436–3440 (2020). https://doi.org/10.18653/v1/2020.coling-main.305
8. Kashefi, O., Hwa, R.: Quantifying the evaluation of heuristic methods for textual data augmentation. In: Proceedings of the Sixth Workshop on Noisy User-generated Text (W-NUT 2020), Online, pp. 200–208 (2020). https://doi.org/10.18653/v1/2020.wnut-1.26
9. Sennrich, R., Haddow, B., Birch, A.: Improving neural machine translation models with monolingual data. In: Proceedings of the 54th Annual Meeting of the Association for Computational Linguistics (Volume 1: Long Papers), Berlin, Germany, pp. 86–96 (2016). https://doi.org/10.18653/v1/P16-1009
10. Sugiyama, A., Yoshinaga, N.: Data augmentation using back-translation for context-aware neural machine translation. In: Proceedings of the Fourth Workshop on Discourse in Machine Translation (DiscoMT 2019), Hong Kong, China, pp. 35–44 (2019). https://doi.org/10.18653/v1/D19-6504
11. Fadaee, M., Bisazza, A., Monz, C.: Data augmentation for low-resource neural machine translation. In: Proceedings of the 55th Annual Meeting of the Association for Computational Linguistics (Vol. 2: Short Papers), Vancouver, Canada, pp. 567–573 (2017). https://doi.org/10.18653/v1/P17-2090

12. Duong, H.-T., Nguyen-Thi, T.-A.: A review: preprocessing techniques and data augmentation for sentiment analysis. Comput. Soc. Netw. **8**(1), 1–16 (2020). https://doi.org/10.1186/s40649-020-00080-x

13. Sahin, G.G., Steedman, M.: Data augmentation via dependency tree morphing for low-resource languages. In: Proceedings of the 2018 Conference on Empirical Methods in Natural Language Processing, Brussels, Belgium, pp. 5004–5009 (2018). https://doi.org/10.18653/v1/D18-1545

Patient Classification Based on Symptoms Using Machine Learning Algorithms Supporting Hospital Admission

Khoa Dang Dang Le[1], Huong Hoang Luong[2], and Hai Thanh Nguyen[3]($\boxtimes$)

[1] Hospital Information System Team, Vietnam Posts and Telecommunications Group, Tien Giang, Vietnam
[2] FPT University, Can Tho, Vietnam
[3] College of Information and Communication Technology, Can Tho University, Can Tho, Vietnam
nthai.cit@ctu.edu.vn

Abstract. Overcrowding in receiving patients, medical examinations, and treatment for hospital admission is common at most hospitals in Vietnam. Receiving and classifying patients is the first step in a medical facility's medical examination and treatment process. Therefore, overcrowding at the regular admission stage has become a complex problem to solve. This work proposes a patient classification scheme representing the text to speed up the patient input flow in hospital admission. First, the Bag of words approach has been built to represent the text as a vector exhibiting the frequency of words in the text. The data used for the evaluation were collected from March 2016 to March 2021 at My Tho City Medical Center - Tien Giang - Vietnam, including 230,479 clinic symptom samples from admissions and discharge office, outpatient department, accident, and Emergency Department. Among learning approaches used in the paper, Logistic Regression reached an accuracy of 79.1% for stratifying patients into ten common diseases in Vietnam. Besides, we have deployed a model explanation technique, Locally Interpretable Model-Agnostic Explanations (LIME), to provide valuable features in disease classification tasks. The experimental results are expected to suggest and classify the patient flow automatically in the hospital admission stage and discharge office to perform the patient flow in the clinics at the hospitals.

Keywords: Hospital admission · Bag of words · Explanation · Clinic symptom · Disease classification

1 Introduction

Receiving and flowing patients into hospital clinics is a rather complicated matter. In [1], the authors stated that "long waiting times became an important

P. Cong Vinh and N. Huu Nhan (Eds.): ICTCC 2021, LNICST 408, pp. 40–50, 2021.
https://doi.org/10.1007/978-3-030-92942-8_4

part of the perception of a general *health care crisis*". For most hospitals in Vietnam, patients have to wait for a long time[1] for hospital admission procedures. Although some solutions have been proposed to reduce and pre-arrange appointments with the doctors'[2], almost all people still tend to crow at the hospital for their health care services. In hospitals, the idea of designing arrangements and suggestions for flowing patients to the medical clinics accurately and quickly helps to increase work efficiency. Usually, in the morning in most hospitals, patients wait for reservations, especially in outpatient departments, accident, and emergency departments, receiving and flowing patients into appropriate hospital clinics to help patients avoid fatigue and minimize conflicts between patients and between patients and medical staff. Receiving patients and flowing patients to clinics in the hospital to diagnose diseases corresponding to signs and clinical symptoms helps avoid wasting time on medical examination and treatment of doctors and patients. Because if the patient's flow is not accurate, the doctor must operate to transfer the patient's room, which is both laborious and time-consuming for the doctor and the patient.

In recent years, the development of machine learning in health care has made significant progress. As a result, machine learning applications have a wide application area, especially supporting and comprehensive care of people's health. For example, numerous applications are related to personal health monitoring [2–8] to reduce disease risk, early detection of modern diseases such as cancer, cardiovascular disease, help reduce costs and prolong life, help patients comply with medication, monitor disease progression. Machine learning is applied in critical areas of medicine: in clinical decisions, electronic health records, diagnostics, medical robots, personalized medicine, medical examination, and treatment management. The progressive application of machine learning in healthcare can improve access to healthcare and the quality of healthcare.

In this work, to develop an automatic patient classification based on symptoms in the text to speed up admission at hospitals, we propose leveraging the Bag of words (BOW) model primarily to generate word features from the text. After converting the text into BOW, we have deployed Term Frequency - Inverse Document Frequency (TF-IDF) [9] to determine the importance of a word in the text to extract the features which are input for machine learning algorithms to perform the classification tasks. Then, we use machine learning algorithms to diagnose ten diseases based on the set of symptoms. After obtaining the learning from models, we have proceeded with an explanation model technique, Local Interpretable Model-agnostic Explanations (LIME) [10], to extract insights in data, including keywords for disease diagnosis. Finally, the experiments are evaluated on the dataset, including 33,024 health record samples at the Health Center of My Tho City - Tien Giang - Vietnam, with ten common diseases in Vietnam.

---

[1] http://baodongnai.com.vn/xahoi/201702/thoi-gian-cho-kham-chua-benh-con-qua-lau-2781882/index.htm, accessed on 01 May 2021.

[2] https://nhandan.vn/tin-tuc-y-te/thi-diem-dat-lich-truc-tuyen-rut-ngan-nua-thoi-gian-kham-chua-benh--643125/, accessed on 10 May 2021.

The rest of this article is as follows. Section 2, we introduce some related work. Section 3, we represent the data for the experiment. Section 4 presents and illustrates our proposed method while Sect. 5 reveals, evaluate the experimental results. The conclusions are rendered in Sect. 6.

2   Related Work

One of the challenges in hospitals today is to make the right decisions to flow patients from the admissions and discharge office to the clinics in the hospital and reduce patient waiting times. Therefore, using machine learning algorithms in the analysis and support of medical staff to perform patient flowing brings more satisfactory results. Furthermore, advances in machine learning algorithms and big data have allowed artificial intelligence to replace humans gradually.

The scientists have proposed a vast amount of research on proposing early diagnosis, improving patient flow in hospitals, and achieving positive results. For example, the study in [11] proposed to improve streaming patients in a hospital emergency department.

Numerous studies have attracted patient classification tasks based on symptoms. In [12], the authors performed quantitative exploration of symptoms using Chi-square test, association with LASSO for analysis to conduct the symptoms of COVID-19 patients. The study in [13] analyzed the Parent-Reported Symptoms to propose efficient treatment for children. The authors in [14] evaluated and experimented on voice features for the Dysphonia-based classification of Parkinson's Disease. The authors in [15] provided a living with uncertainty method that mapped the transition from pre-diagnosis to a diagnosis of dementia. The work in [16] deployed the (TF/IDF), Bag of words (BOW) to analyze symptoms on COVID-19 clinical text data. Finally, the work in [17] applied standard text categorization methods for exploring patient text. The proposed aimed to predict treatment outcomes in Internet-based cognitive-behavioral therapy.

With the BOW and TF/IDF achievements, our work has deployed these approaches to analyze the patient clinical symptoms of 10 diseases in the text at the Health Center of My Tho City - Tien Giang, Vietnam. We combine the obtained vector with meaningful attributes in the patient's medical examination and treatment data to build a model to classify patients into common diseases for hospital admission.

3   Dataset Description

This research examined the PatientAdmission dataset, with 230,479 samples containing the information: age, gender, clinical symptoms of the patient to propose building a machine learning model, which supports the implementation of automatic classification diseases and decision to flow patients to hospital clinics. Data were collected from March 2016 to March 2021 from the Medical Center of My Tho City - Tien Giang - VietNam, from the admissions and discharge office, outpatient department, accident, emergency department, and related reports.

Table 1. Information on the considered dataset with the number of samples

No.	Disease name	Total of samples
1	Neoplasms	16271
2	Endocrine, Nutritional and metabolic diseases	38672
3	Diseases of the eye and adnexa	18443
4	Diseases of the circulatory system	37782
5	Diseases of the respiratory system	41888
6	Diseases of skin and subcutaneous tissue	7044
7	Diseases of the musculoskeletal system and connective tissue	35427
8	Diseases of the genitourinary system B212	17503
9	Pregnancy, childbirth and the puerperium	3666
10	Injury, poisoning and certain other consequences of external causes	13783

Data are manually or semi-automatically retrieved patient information (the QRCode in the health insurance card). The dataset includes patient's age (AGE field), patient's gender 1 Male, 0 Female (SEX field), and CLINICAL SYMPTOMS field. Clinical symptoms were noted by the medical staff who received the patient when the patient was declared at the admission and discharge office. The patient's clinical symptom data may contain information about physical fitness, abnormal vital signs, symptoms, and manifestations before and during hospital arrival. ID_DISEASES field is about a patient's type disease that is recognized after being examined, treated by a doctor, and identified with the patient's ICD10- disease code, a group of diseases that includes many diseases, a type disease that includes many groups of diseases. In the table of data, there are ten considered common type diseases which were recorded in Vietnam hospitals (exhibited in Table 1).

4   Methods

The overall architecture for proposed hospital admissions is exhibited in Fig. 1. The data are extracted from medical information systems[3] at the Health Center of My Tho City - Tien Giang, Vietnam and divided into training and test set. Feature engineering is performed with the BOW technique on the training set to provide a word vector containing a dictionary that includes words extracted from clinical symptoms and signs described in pre-diagnosis records when the patients are admitted into the hospital. Then, we use the TF-IDF [9] method to calculate the value for features which are words included in the BOW extracted from the training data. Finally, we use the training set is fetched into learning algorithms to build a model for evaluating the test set. Details of methods are introduced in the following sections.

---

[3] https://hotrovnpthis.wordpress.com/, accessed on 01 May 2021.

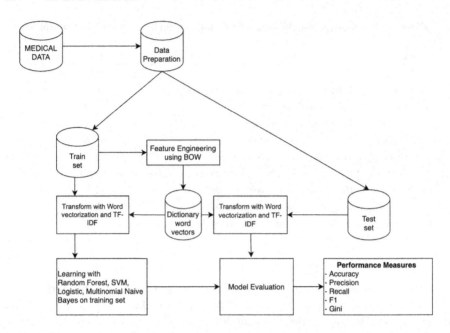

Fig. 1. The proposed architecture for patient classification based on symptoms description

The BOW model is a wording technique implemented in natural language processing. More specifically, text representation is the Bag of words. In this model, BOW disregarding grammar but keeping multiplicity [1]. Let us consider the following example:

(1) Jong has a bad headache. Ronaldo also has a bad headache.
(2) Nadal is going to the doctor for a headache

Considering these texts, we see that a list of words is generated as follows for each text:

"Jong ", "has ","a", "bad", "headache", "Ronaldo ", "also"

"Nadal","is", "going", "to", "the", "doctor", "for", "a", "headache"

Representing each BOW as a JSON object as follows:

B1 = "Jong":1,"has":2,"a":2,"bad":2,"headache":2,"Ronaldo":1,"also":1

B2 = "Nadal":1,"is":1,"going":1,"to":1,"the":1,"doctor":1,"for":1,"a":1,
"headache":1

In order to represent text, each word can be represented as the key of the BOW model, while it corresponds to the value of the number of occurrences and the order of the words that can be ignored:

"Jong":1,"has":2,"Ronaldo":1,"also":1,"a":2,"bad":2,"headache":2 also B1.

When the dataset is a few hundred documents, we can see that the dictionary can range from a few tens of thousands of words to a few hundred thousand words. Therefore, the number of dimensions obtained after transforming BOW is vast. Some machine learning models like Naive Bayes handle it inefficiently. To overcome, we can use the method of reducing the number of data dimensions. The reduction method can select the most important words to distinguish one text from another or the dimensionality reduction method. However, this reduction step often causes information loss, reducing the accuracy of the classifier after implementation.

We transformed the text with Term Frequency - Inverse Document Frequency (TF-IDF) which is considered as a robust numerical statistic tool to reveal essential words in a text document[9].

After applying such these approaches with BOW and TF-IDF, we fetch transformed into learning algorithms such as Random Forest, Support Vector Machines, Logistic Regression, and Multinomial Naive Bayes with hyper-parameters as presented in Table 2. In comparison with various learning methods, we use accuracy, precision, recall, f1-score, and Gini score.

Table 2. Information of hyper-parameters use in the training phase

Machine learning algorithm	Hyper-parameters
Logistic Regression	C: 1.0, penalty: l1
SVM	C: 10, gama: 0.0001, kernel: rbf
Decision Tree	Criterion: gini, max depth: 450
Multinomial Naive Bayes	Default value: Laplace/Lidstone smoothing value of 1.0
Random Forest	Criterion: gini, max depth: 450, max features: auto, 750 trees

5 Experimental Results

5.1 Setting

This study used a PC Windows 10 Home Single Language 64 bit with a CPU of 28 cores Intel(R) Xeon(R) CPU E5 - 2680 v4 @ 2.4 GHz and 96 GB of RAM that is a sufficiently responsive hardware environment to test the proposed algorithm. The experiment code for all the models is available through python that supports several libraries, such as Sklearn [18], Matplotlib [19], and Pandas [20,21].

5.2 Disease Classification Task with Various Algorithms

In Table 3, we compare performance among considered classifiers. As observed, Logistic Regression reveals the best result while Multinomial Naive Bayes exhibits the worst one.

In the above analysis, using the Logistic Regression algorithm to achieve the best results with measures Accuracy = 79.1%, Precision = 79.9%, Recall = 79.1%, F1 = 79.5%, Gini = 92.1% and time 159 s. Multinomial Naive Bayes algorithm

Table 3. The disease classification performance performed by various learning approaches. The **bold** results are the best results among the considered learning algorithms

Method	ACC (%)	Precision (%)	Recall (%)	F1 (%)	Gini(%)	Time (s)
RandomForest	78.6	79.3	78.6	78.9	89.7	9009
SVM	78.0	78.7	78.0	78.3	88.8	237
Decision Tree	74.0	74.9	74.0	74.4	70.6	481
Multinomial Naive Bayes	72.7	75.1	72.7	73.9	88.7	46
Logistic Regression	**79.1**	**79.9**	**79.1**	**79.5**	**92.1**	159

achieved the lowest results with the measures Accuracy = 72.7%, Precision = 75.1%, Recall = 72.7%, F1 = 72.5%, Gini = 88.7% and time 246 s.

The algorithms SVM, Random Forest gives results with a reasonably high Accuracy measure of more than 78%. The algorithm's Decision Tree gives results with the measures accuracy = 74%.

5.3 Analysis of Disease Symptoms

Some interesting analyzes from the dataset when using the LIME technique to analyze important words affecting type disease classification performed by Logistic Regression shown in Figs. 2 and 3.

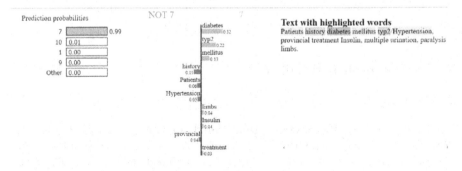

Fig. 2. Use LIME to Explain the prediction of Logistic Regression and display the words symptoms description that can be key word to describe the disease

We extract the essential features in the above classification model, important information that dramatically affects the patient classification process is shown in the patient's blood pressure and temperature measurements, this shows the possibility of when measuring body temperature, measuring blood pressure when asking the patient at the admissions and discharge office, the model predicts the possibility of having the disease.

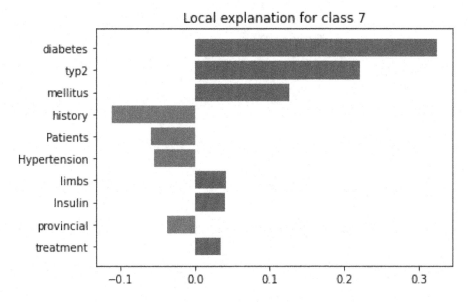

Fig. 3. Use LIME to Explain compare the importance of words affecting data

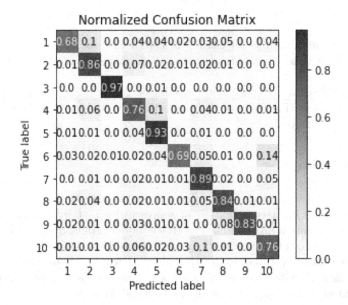

Fig. 4. Confusion matrix of prediction results with Logistic Regression

We reveal the confusion matrix in Fig. 4 of Logistic Regression, where we obtain the best results. Because we have an imbalanced dataset, the classes are not represented equally. The model is perfect for classes: 3 and 5 with more than

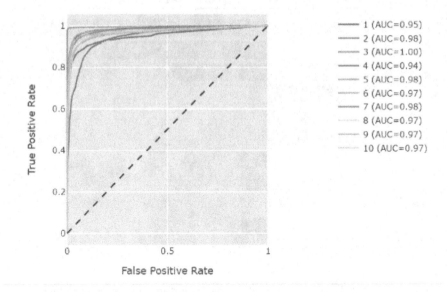

Fig. 5. ROC curve

90% accuracy, and suitable for classes: 2, 7, 8, 9 with more than 80% accuracy, not suitable for class 1 with 68% accuracy (the indexes of classes correspond to No. in Table 1).

Figure 5 illustrates experimental results of Logistic Regression in another metric, ROC curve. The curve of 10 diseases is above the diagonal, demonstrating that the model is good.

6   Conclusion

We have presented a method to solve automatic patient input flowing. Our research is based on a combination of representing text by the BOW model and data classification algorithms. The BOW model is built quickly to represent the text as a vector of occurrence frequency of words in the text with the number of dimensions is very large. Incorporating other patient-specific information allows efficient classification of this data set. The proposed approach provides an automated way of patient flow in the hospital to improve healthcare in the future.

Acknowledgement. To conduct this study, our team would like to express our gratitude to the Hospital Information System team in Vietnam Posts and Telecommunications Group, Tien Giang province, Vietnam, for providing valuable data for our work.

References

1. Ringard, Å., Hagen, T.P.: Are waiting times for hospital admissions affected by patients' choices and mobility? BMC Health Serv. Res. **11**(1) (2011). https://doi.org/10.1186/1472-6963-11-170
2. Sajedi, S.O., Liang, X.: Uncertainty-assisted deep vision structural health monitoring. Comput.-Aided Civ. Infrastruct. Eng. **36**(2), 126–142 (2020). https://doi.org/10.1111/mice.12580
3. Valsalan, P., Baomar, T.A.B., Baabood, A.H.O.: IOT based health monitoring system. J. Critic. Rev. **7**(04), 739–743 (2020). https://doi.org/10.31838/jcr.07.04.137
4. Dong, C.Z., Catbas, F.N.: A review of computer vision-based structural health monitoring at local and global levels. Struct. Health Monit. **20**(2), 692–743 (2020). https://doi.org/10.1177/1475921720935585
5. Nasiri, S., Khosravani, M.R.: Progress and challenges in fabrication of wearable sensors for health monitoring. Sens. Actuat. A Phys. **312**, 112105 (2020), https://doi.org/10.1016/j.sna.2020.112105
6. Li, C., Sun, L., Xu, Z., Wu, X., Liang, T., Shi, W.: Experimental investigation and error analysis of high precision FBG displacement sensor for structural health monitoring. Int. J. Struct. Stab. Dyn. **20**(06), 2040011 (2020). https://doi.org/10.1142/s0219455420400118
7. Kim, J., et al.: Self-charging wearables for continuous health monitoring. Nano Energy **79**, 105419 (2021), https://doi.org/10.1016/j.nanoen.2020.105419
8. Chen, Z., Sheng, H., Xia, Y., Wang, W., He, J.: A comprehensive review on blade tip timing-based health monitoring: status and future. Mech. Syst. Sig. Process. **149**, 107330 (2021), https://doi.org/10.1016/j.ymssp.2020.107330
9. Uther, W., et al.: TF-IDF. In: Encyclopedia of Machine Learning, pp. 986–987. Springer, Boston (2011). https://doi.org/10.1007/978-0-387-30164-8_832
10. Ribeiro, M.T., Singh, S., Guestrin, C.: Why should i trust you?: explaining the predictions of any classifier. In: Proceedings of the 2016 Conference of the North American Chapter of the Association for Computational Linguistics: Demonstrations (2016)
11. Medeiros, D.J., Swenson, E., DeFlitch, C.: Improving patient flow in a hospital emergency department. In: 2008 Winter Simulation Conference, pp. 1526–1531 (2008)
12. Qu, G., et al.: A quantitative exploration of symptoms in COVID-19 patients: an observational cohort study. Int. J. Med. Sci. **18**(4), 1082–1095 (2021). https://doi.org/10.7150/ijms.53596
13. Molloy, M.A., et al.: Parent-reported symptoms and perceived effectiveness of treatment in children hospitalized with advanced heart disease. J. Pediatr. (2021). https://doi.org/10.1016/j.jpeds.2021.06.077
14. Goyal, J., Khandnor, P., Aseri, T.C.: A comparative analysis of machine learning classifiers for dysphonia-based classification of parkinson's disease. Int. J. Data Sci. Anal. **11**(1), 69–83 (2020). https://doi.org/10.1007/s41060-020-00234-0
15. Campbell, S., et al.: Living with uncertainty: mapping the transition from pre-diagnosis to a diagnosis of dementia. J. Aging Stud. **37**, 40–47 (2016). https://doi.org/10.1016/j.jaging.2016.03.001
16. Khanday, A.M.U.D., Rabani, S.T., Khan, Q.R., Rouf, N., Mohi Ud Din, M.: Machine learning based approaches for detecting COVID-19 using clinical text data. Int. J. Inf. Technol. **12**(3), 731–739 (2020). https://doi.org/10.1007/s41870-020-00495-9

17. Gogoulou, E., Boman, M., Ben Abdesslem, F., Hentati Isacsson, N., Kaldo, V., Sahlgren, M.: Predicting treatment outcome from patient texts: the case of Internet-based cognitive behavioural therapy. In: Proceedings of the 16th Conference of the European Chapter of the Association for Computational Linguistics: Main Volume. pp. 575–580. Association for Computational Linguistics, April 2021. https://aclanthology.org/2021.eacl-main.46
18. Pedregosa, F., et al.: Scikit-learn: machine learning in Python. J. Mach. Learn. Res. **12**, 2825–2830 (2011)
19. Hunter, J.D.: Matplotlib: a 2D graphics environment. Comput. Sci. Eng. **9**(3), 90–95 (2007)
20. Pandas Development Team: pandas-dev/pandas: Pandas, February 2020. https://doi.org/10.5281/zenodo.3509134
21. Wes McKinney: Data structures for statistical computing in python. In: van der Walt, S., Millman, J. (eds.) Proceedings of the 9th Python in Science Conference, pp. 56–61 (2010)

Implementation of the Motion Controller for an Omni Mobile Robot Using the Trajectory Linearization Control

Vu Hoan Bui[1,2], Chi Tuan Duong[1,2], Trong Nhan Nguyen[1,2],
The Cuong Le[1,2], Xuan Quang Ngo[1,2], Duy Khang Luong[1,2],
Nhat An Ho[1,2], Huy Vu Phung[1,2], Tri Manh Vo[1,2],
and Ha Quang Thinh Ngo[1,2(✉)]

[1] Department of Mechatronics Engineering, Faculty of Mechanical Engineering,
Ho Chi Minh City University of Technology (HCMUT), 268 Ly Thuong Kiet
Street, District 10, Ho Chi Minh City 700000, Vietnam
nhqthinh@hcmut.edu.vn
[2] Vietnam National University Ho Chi Minh City, Linh Trung Ward, Thu Duc
District, Ho Chi Minh City 700000, Vietnam

Abstract. This study introduces the design and modeling of nonlinear controller based on the omni-wheel for mobile robot. Via/Through the dynamics characteristics, the control scheme of Trajectory Linearization Control (TLC) is developed. Furthermore, the proposed controller is simulated in order to verify the effectiveness and feasibility. From the simulation results, the proposed controller achieves superior performance such 20% better tracking error, 10% faster transient period and converge to finite time.

Keywords: Omni-robot · Mobile robot · Trajectory linearization control · Simulation

1 Introduction

The main object of this research is an omni-directional robot which is a kind of holonomic mobile platform. The flexibility, maneuverability and highly practical application capabilities cause why many scientists have researched about this robot [1].

In our investigation, omni-directional robot has 03 orthogonal wheels that are fixed at each vertex of an even triangle and controlled by the DC servo motors and the main MCU (micro-controller unit). Also, an embedded computer is utilized to calculate the inverse kinetic and receive directives from main PC (personal computer) through LAN (local area network). Besides, the position of the object is determined by a roof camera. Base on this location, PC could manipulate the mission for tracking trajectory to follow the path that is generated by the process of trajectory planning.

With the above researches, most of previous studies have been targeted on the dynamic analysis and mechanical design. However, the dynamical model and highly precise control to track path/trajectory have not been studied deeply. In many topics of omni-directional dynamic models, the ideal servo motor is used to drive the motors as

P. Cong Vinh and N. Huu Nhan (Eds.): ICTCC 2021, LNICST 408, pp. 51–66, 2021.
https://doi.org/10.1007/978-3-030-92942-8_5

well as the responses of motor can adapts to the desired command without error [7]. In the real condition, the dynamical constraints of servo motor could significantly affect on the robot motion. The dynamic model and control strategies of the omni-directional mobile robot is developed and discussed in [2] and [3]. Nevertheless, these controllers are based on the linear control method while the robot dynamics are nonlinear. Additionally, the feedback gain of the controllers has to be tuned to achieve system stability for different trajectories. This is also solved by fuzzy and adaptive method in [3]. But in a dynamic environment such as sportive applications, it is really difficult to predetermine the trajectory and adjust the feedback gain in real-time to combine both stability and transient response of the robot control system. Besides, [8] introduces an adaptive control method unknowing the center of gravity (C.o.G). However, in our scope, C.o.G could be determined via mechanical design software and considered unchanged during working time. Friction and slip are also ignored despite being mentioned at [9] and handled by an adaptive backstepping control method. Additionally, the feedback gain of the controllers has to be tuned to achieve system stability for different trajectories.

In the dynamic environment such sportive applications, it is really difficult to predetermine the trajectory and adjust the feedback gain in real time to combine both stability and transient response of the robot control system. To achieve better performance, it is desired that the controller for tracking trajectory could follow any given feasible path. In this research, TLC scheme on omni-directional mobile robot which combines nonlinear inverse kinetic duty of linearly time-varying eigen structure, is investigated [4]. The nonlinear tracking and decoupling control by trajectory linearization can be handled as the perfect gain-scheduling controller that designed at every point on the trajectory. Therefore, the TLC provides robust stability and excellent performance along the trajectory without interpolation of controller gains.

In Sect. 2: The kinematics, dynamics of the robot and the DC motor of omni-directional mobile robot model is presented.

In Sect. 3: Create controller for the robot using TLC method.

In Sect. 4: Simulation test result.

In Sect. 5: Conclusion & Plan.

2   The Mathematical Model of Omni-Directional Mobile Robot

Initially, in order to design a controller for robot, the descriptions of robot motion are essential to be indicated with some assumptions as below.

Assumption 1: Since the electrical time constant of the motor is too small, the wheels roll on a flat without slipping and the friction force is simplified to be represented by a viscous damping coefficient.

Assumption 2: The unmodeled dynamics will be compensated by the feedback controller which is based on this simplified model in the next section.

2.1   Coordinates and Symbols

Let the mobile robot be rigid moving on the workspace. It is assumed that the world coordinate system $O_w X_w Y_w Z_w$ is fixed on the ground, and the local coordinate system $O_m X_m Y_m Z_m$ is fixed on the C.o.G (Center of Gravity) of the mobile robot as Fig. 1.

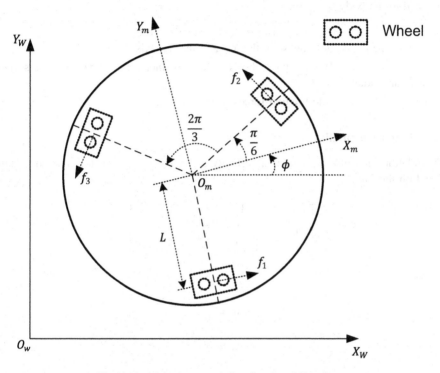

Fig. 1. Model of an omni-directional mobile robot

The following mathematical notations are utilized to depict this system:

- L: the distance between driving motor and C.o.G of the robot
- m: the weight of robot
- I: the moment inertia of robot
- R: the radius of driving wheels
- n: gear ratio between motor and wheel
- $w = [w_1 \quad w_2 \quad w_3]^T$: the angular velocities of the 1st, 2nd and 3rd wheel respectively
- $w_m = [w_{m1} \quad w_{m2} \quad w_{m3}]^T$: the angular velocities of the 1st, 2nd and 3rd motor correspondingly
- $v_T = [v_{w1} \quad v_{w2} \quad v_{w3}]^T$: the tangential velocity of the 1st, 2nd and 3rd wheel respectively

Then, the location of robot is expressed in the world coordinate as following

- $^Wr = \begin{bmatrix} ^Wx & ^Wy & ^W\phi \end{bmatrix}^T$: the position of the robot's C.o.G
- $^Wv = \begin{bmatrix} ^Wv_x & ^Wv_y & ^Wv_\phi \end{bmatrix}^T$: the robot's velocities consisting of linear velocities and angular velocity

Moreover, some system parameters are illustrated in the local coordinate which attached to the body of robot.

- $^mr = \begin{bmatrix} ^mx & ^my & ^m\phi \end{bmatrix}^T$: the position of the robot's C.o.G
- ϕ: the rotational angle between the X_W and the X_m axis
- $^mv = \begin{bmatrix} ^mv_x & ^mv_y & ^mv_\phi \end{bmatrix}^T$: the robot's velocities including linear velocities and angular velocity
- $f = \begin{bmatrix} f_1 & f_2 & f_3 \end{bmatrix}^T$: the traction forces of the wheels.

2.2 Robot Kinematics

The relationship between tangential velocity at each wheel and velocity of the robot based on the Fig. 2.

$$\begin{bmatrix} v_{w1} \\ v_{w2} \\ v_{w3} \end{bmatrix} = \begin{bmatrix} 1 & 0 & L \\ -\frac{1}{2} & \frac{\sqrt{3}}{2} & L \\ -\frac{1}{2} & -\frac{\sqrt{3}}{2} & L \end{bmatrix} \begin{bmatrix} ^mv_x \\ ^mv_x \\ ^mv_\phi \end{bmatrix} \tag{1}$$

We have,

$$w = \frac{v_T}{R} \tag{2}$$

$$w_m = nw \tag{3}$$

Since the driving motor is directly attached to the omni wheel, from Eq. (1), (2) and (3) the relationship between the rotational velocities of the motors and its velocities is determined as

$$\begin{bmatrix} w_{m1} \\ w_{m2} \\ w_{m3} \end{bmatrix} = B^T \frac{n}{R} \begin{bmatrix} ^mv_x \\ ^mv_y \\ ^mv_\phi \end{bmatrix} \tag{4}$$

where, $B = \begin{bmatrix} 1 & -\frac{1}{2} & -\frac{1}{2} \\ 0 & \frac{\sqrt{3}}{2} & -\frac{\sqrt{3}}{2} \\ L & L & L \end{bmatrix}$

As a result, with the matrix transformation ^WR_m from the local coordinate to the world coordinate, the kinematic equation of the system could be re-written as

$$^Wv = {}^WR_m{}^mv = {}^WR_m \frac{R}{n} (B^T)^{-1} w_m \tag{5}$$

Where, $^{W}R_m = \begin{bmatrix} \cos\phi & -\sin\phi & 0 \\ \sin\phi & \cos\phi & 0 \\ 0 & 0 & 1 \end{bmatrix}$.

2.3 Robot Dynamic

The acceleration of the point with respect to the world coordinate system $O_w X_w Y_w Z_w$

$$\vec{a}_{P/O_w} = \vec{a}_{O_m/O_w} + \left(\vec{a}_{P/O_m}\right)_{O_m X_m Y_m Z_m} + 2\vec{w}_{O_w} \times \left(\vec{v}_{P/O_m}\right)_{O_m X_m Y_m Z_m} \\ + \dot{\vec{w}} \times \vec{r}_{P/O_m} + \vec{w}_{O_w} \times \vec{v}_{P/O_m}$$ (6)

Where,

- $\vec{a}_{P/O_w}$: the acceleration of point P, the point on the robot, with respect to the frame O_w;
- $\vec{a}_{O_m/O_w}$: the acceleration of the frame O_m with respect to the frame O_w; $\left(\vec{a}_{P/O_m}\right)_{O_m X_m Y_m Z_m}$: the acceleration of point P with respect to the frame O_m, $\left(\vec{a}_{P/O_m}\right)_{O_m X_m Y_m Z_m} = 0$;
- $2\vec{w}_{O_w} \times \left(\vec{v}_{P/O_m}\right)_{O_m X_m Y_m Z_m}$: the Coriolis acceleration, $2\vec{w}_{O_w} \times \left(\vec{v}_{P/O_m}\right)_{O_m X_m Y_m Z_m} = 0$;
- $\dot{\vec{w}} \times \vec{r}_{P/O_m}$: the Euler acceleration, $\dot{\vec{w}} \times \vec{r}_{P/O_m} = 0$;
- $\vec{w}_{O_w} \times \vec{v}_{P/O_m}$: the centripetal acceleration;
- $\dot{\vec{w}} = \dot{\phi}\,\vec{k}$;
- $\vec{v}_{P/O_m} = {}^m v_x\,\vec{i} + {}^m v_y\,\vec{j}$;
- $\vec{w} \times \vec{v}_{P/O_m} = r\vec{k} \times \left({}^m v_x\,\vec{i} + {}^m v_y\,\vec{j}\right) = r\,{}^m v_x\,\vec{j} - r\,{}^m v_y\,\vec{i}$;

So, rewrite the Eq. (6)

$$\vec{a}_{P/O_w} = \vec{a}_{O_m/O_w} + r\,{}^m v_x\,\vec{j} - r\,{}^m v_y\,\vec{i}$$ (7)

Applying Newton's law

$$\sum \vec{F} = m\vec{a}_{P/O_w}$$ (8)

Projecting the Eq. (8) onto $O_m X_m$, $O_m Y_m$, $O_m Z_m$ axis respectively

$$F_X = m\left({}^m \dot{v}_x - r\,{}^m v_y\right)$$ (9)

$$F_Y = m\left({}^m \dot{v}_y + r\,{}^m v_x\right)$$ (10)

$$T = I^m \dot{v}_\phi \tag{11}$$

Obtaining the dynamical performance of the omni mobile robot based on the Eq. (9), (10) and (11), we have

$$\begin{bmatrix} {}^m\dot{v}_x \\ {}^m\dot{v}_y \\ {}^m\dot{v}_\phi \end{bmatrix} = \begin{bmatrix} {}^mv_\phi {}^mv_y \\ -{}^mv_\phi {}^mv_x \\ 0 \end{bmatrix} + HB \begin{bmatrix} f_1 \\ f_2 \\ f_3 \end{bmatrix} \tag{12}$$

Where, $H = \begin{bmatrix} \frac{1}{m} & 0 & 0 \\ 0 & \frac{1}{m} & 0 \\ 0 & 0 & \frac{1}{I} \end{bmatrix}$

In [5], the dynamic characteristics of each DC motor can be demonstrated as

$$L_a \frac{di_a}{dt} + R_a i_a + k_b w_m = u \tag{13}$$

$$J_m \dot{w}_m + D_m w_m + \frac{Rf}{n} = k_t i_a \tag{14}$$

Where,

- u is the applied armature voltage;
- i_a is the armature current;
- L_a is the armature inductance;
- R_a is the armature resistance;
- k_b is the back emf constant;
- k_t is the motor torque constant;
- J_m is the equivalent inertia at the armature including both the armature inertia and the load inertia referred to the armature such as gear train, wheel, etc.;
- D_m is the equivalent viscous damping at the armature including both the armature viscous damping and the load viscous damping referred to the armature such as gear train, wheel, etc.

Since the electrical time constant is negligible comparing to the mechanical time constant, we could achieve

$$\frac{di_a}{dt} = 0 \tag{15}$$

Substituting the Eq. (15) into the Eq. (14), we have

$$i_a = \frac{1}{R_a}(u - k_b w_m) \tag{16}$$

From the Eqs. (4), (12), (14) and (16), we get the dynamics equation of the system in the body frame in respect to the supplied voltage to the motor $u = \begin{bmatrix} u_1 & u_2 & u_3 \end{bmatrix}^T$

$$
G \begin{bmatrix} {}^m\dot{v}_x \\ {}^m\dot{v}_y \\ {}^m\dot{v}_\phi \end{bmatrix} = \begin{bmatrix} {}^m v_\phi {}^m v_y \\ -{}^m v_\phi {}^m v_x \\ 0 \end{bmatrix} - HBB^T \frac{n^2}{R^2} \left(D_m + \frac{k_t k_b}{R_a} \right) \begin{bmatrix} {}^m v_x \\ {}^m v_y \\ {}^m v_\phi \end{bmatrix}
$$
$$
+ HB \frac{n}{R} \frac{k_t}{R_a} \begin{bmatrix} u_1 \\ u_2 \\ u_3 \end{bmatrix}
\tag{17}
$$

Where, $G = I + HBB^T J_m \frac{n^2}{R^2}$.

3 Controller Design

In the previous works [6], three speed controllers of the driving motor were independently installed in omni-directional mobile robot. Like almost omni-directional robots, the signal control of each motor controller is established by the inverse dynamic of the robot model. In fact, most of mobile robots could not track the desired following path because the open-loop controller exists inevitable errors. The errors should be eliminated or reduced to zero.

Designing a controller design bases on TLC, which is showed in this section. TLC has been proven to be a successful technique in nonlinear decoupling, disturbance reduction and following path [2, 4, 7]. TLC is considered as the ideal gain-scheduling controller to track point on the path. The controller consists of two loop parts. The inside loop is the hardware constraints of controller to obey the control command from the outside loop. The outside loop modifies the position of robot based on the information of the vision system. The first part is the feedforward controller, which is formulated to automatically tuning the parameters of a system by mean of dynamic inversion. The second part is a feedback controller, which forces the robot to return to the reference trajectory.

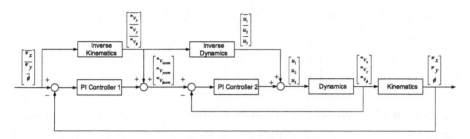

Fig. 2. The structure of proposed controller using TLC for the omni mobile robot

3.1   Outer Loop Control

Through the kinematic Eq. (5), based on the desired trajectory $\begin{bmatrix} \overline{W}_x(t) & \overline{W}_y(t) & \overline{\phi}(t) \end{bmatrix}^T$, the nominal velocity of the system might be calculated as

$$
\begin{bmatrix} \overline{\dot{m}x} \\ \overline{\dot{m}y} \\ \dot{\overline{\phi}} \end{bmatrix} = \begin{bmatrix} \cos\overline{\phi} & \sin\overline{\phi} & 0 \\ -\sin\overline{\phi} & \cos\overline{\phi} & 0 \\ 0 & 0 & 1 \end{bmatrix} \begin{bmatrix} \overline{\dot{W}x} \\ \overline{\dot{W}y} \\ \dot{\overline{\phi}} \end{bmatrix} \tag{18}
$$

The linearization of the function about the equilibrium point is then given by:

$$
\dot{x}_i = f_i(x_1, x_2, ..., x_n, u_1, u_2, ..., u_m)
$$
$$
\approx \sum_{j=1}^{n} \frac{\partial f_i}{\partial x_j}\Big|_{x_j = \overline{x}_j} (x_j - \overline{x}_j) + \sum_{j=1}^{m} \frac{\partial f_i}{\partial u_j}\Big|_{u_j = \overline{u}_j} (u_j - \overline{u}_j) \tag{19}
$$

Where,

- $x_1, x_2, ..., x_n, u_1, u_2, ..., u_m$: the states and inputs;
- $\overline{x}_1, \overline{x}_2, ..., \overline{x}_n, \overline{u}_1, \overline{u}_2, ..., \overline{u}_m$: the equilibrium points;

From the Eq. (18) and the Eq. (19), we have

$$
\dot{x}_1 = {}^W\dot{x} = \cos\phi\, {}^mv_x - \sin\phi\, {}^mv_y = f_1\left({}^Wx, {}^Wy, \phi, {}^mv_x, {}^mv_y, \phi\right) \tag{20}
$$

$$
\dot{x}_2 = {}^W\dot{y} = \cos\phi\, {}^mv_x - \sin\phi\, {}^mv_y = f_2\left({}^Wx, {}^Wy, \phi, {}^mv_x, {}^mv_y, \phi\right) \tag{21}
$$

$$
\dot{x}_3 = \dot{\phi} = f_3\left({}^Wx, {}^Wy, \phi, {}^mv_x, {}^mv_y, \phi\right) \tag{22}
$$

Furthermore, we define several mathematical notations for tracking purpose

$$
\begin{bmatrix} e_x \\ e_y \\ e_\phi \end{bmatrix} = \begin{bmatrix} {}^Wx \\ {}^Wy \\ \phi \end{bmatrix} - \begin{bmatrix} \overline{W}_x \\ \overline{W}_x \\ \overline{\phi} \end{bmatrix} \tag{23}
$$

$$
\begin{bmatrix} \delta^m v_x \\ \delta^m v_y \\ \delta^m v_\phi \end{bmatrix} = \begin{bmatrix} {}^mv_{xcom} \\ {}^mv_{ycom} \\ {}^mv_{\phi com} \end{bmatrix} - \begin{bmatrix} \overline{{}^mv_x} \\ \overline{{}^mv_y} \\ \overline{{}^mv_\phi} \end{bmatrix} \tag{24}
$$

Linearizing Eq. (20), (21), (22) about the nominal trajectory $\begin{bmatrix} \overline{W}_x(t) & \overline{W}_y(t) & \overline{\phi}(t) \end{bmatrix}^T$ and $\begin{bmatrix} \overline{{}^mv_x} & \overline{{}^mv_y} & \overline{{}^mv_\phi} \end{bmatrix}^T$ based on the Eq. (19), we have

$$
\dot{e}_x = \left(-\sin\overline{\phi}\, {}^mv_x - \cos\overline{\phi}\, {}^mv_y\right)e_\phi + \cos\phi\,\delta^m v_x - \sin\phi\,\delta^m v_y \tag{25}
$$

$$\dot{e}_y = \left(\cos \overline{\phi}^m v_x - \sin \overline{\phi}^m v_y\right) e_\phi + \sin \phi \delta^m v_x + \cos \phi \delta^m v_y \tag{26}$$

$$\dot{e}_\phi = \delta^m v_\phi \tag{27}$$

From Eq. (25), (26) and (27), we have

$$\begin{bmatrix} \dot{e}_x \\ \dot{e}_y \\ \dot{e}_\phi \end{bmatrix} = A_1 \begin{bmatrix} e_x \\ e_y \\ e_\phi \end{bmatrix} + B_1 \begin{bmatrix} \delta^m v_x \\ \delta^m v_y \\ \delta^m v_\phi \end{bmatrix} \tag{28}$$

where,

$$A_1 = \begin{bmatrix} 0 & 0 & -\sin \overline{\phi}^m v_x - \cos \overline{\phi}^m v_y \\ 0 & 0 & \cos \overline{\phi}^m v_x - \sin \overline{\phi}^m v_y \\ 0 & 0 & 0 \end{bmatrix}; \quad B_1 = \begin{bmatrix} \cos(\phi) & -\sin(\phi) & 0 \\ \sin(\phi) & \cos(\phi) & 0 \\ 0 & 0 & 1 \end{bmatrix}$$

Later, the PI (Proportional-Integral) controller is designed to stabilize the system performance

$$\begin{bmatrix} \delta^m v_x \\ \delta^m v_y \\ \delta^m v_\phi \end{bmatrix} = -K_{P1} \begin{bmatrix} e_x \\ e_y \\ e_\phi \end{bmatrix} - K_{I1} \begin{bmatrix} \int e_x(t) dt \\ \int e_y(t) dt \\ \int e_\phi(t) dt \end{bmatrix} \tag{29}$$

From the Eq. (28) and (29), we get

$$\dot{x}_o = A_{clo} x_o \tag{30}$$

Where,

$$x_o = \begin{bmatrix} \int e_x(t) dt \\ \int e_y(t) dt \\ \int e_\phi(t) dt \\ e_x(t) \\ e_y(t) \\ e_\phi(t) \end{bmatrix}; \quad A_{clo} = \begin{bmatrix} O_3 & I_3 \\ -B_1 K_{I1} & A_1 - B_1 K_{P1} \end{bmatrix}; \\ O_3 \text{ is the } 3 \times 3 \text{ zero matrix;} \\ I_3 \text{ is the } 3 \times 3 \text{ identity matrix;} \tag{31}$$

Design K_{P1} and K_{I1} to achieve the desired closed-loop tracking error dynamics

$$A_{clo} = \begin{bmatrix} O_3 & I_3 \\ diag[-a_{111} & -a_{121} & -a_{131}] & diag[-a_{112} & -a_{122} & -a_{132}] \end{bmatrix} \tag{32}$$

Where $a_{1j1} > 0$, $a_{1j2} > 0$, $j = 1, 2, 3$ are the coefficients of the desired closed-loop characteristic polynomial of each channel.

Based on the desired response and tracking error, we choose the suitable poles, and then define the coefficients of the desired closed-loop characteristic polynomial.

From Eq. (31) and Eq. (32), we get the gain K_{P1} and K_{I1} of the controller

$$
\begin{aligned}
K_{I1} &= -B_1^{-1} diag[\, -a_{111} \quad -a_{121} \quad -a_{131} \,] \\
K_{P1} &= B_1^{-1}(A_1 - diag[\, -a_{112} \quad -a_{122} \quad -a_{132} \,])
\end{aligned}
\tag{33}
$$

The tracking command for the inner loop is given by

$$
\begin{bmatrix} {}^m v_{xcom} \\ {}^m v_{ycom} \\ {}^m v_{\phi com} \end{bmatrix} = \begin{bmatrix} \overline{{}^m v_x} \\ \overline{{}^m v_y} \\ \overline{{}^m v_\phi} \end{bmatrix} + \begin{bmatrix} \delta^m v_x \\ \delta^m v_y \\ \delta^m v_\phi \end{bmatrix}
\tag{34}
$$

3.2   Inner Loop Control

By using the kinematic Eq. (17), we can get the nominal values of control voltages for the driving motors owing to the nominal velocities of the system $\begin{bmatrix} \overline{{}^m v_x} & \overline{{}^m v_y} & \overline{{}^m v_\phi} \end{bmatrix}^T$

$$
\begin{aligned}
\begin{bmatrix} \overline{u_1} \\ \overline{u_2} \\ \overline{u_3} \end{bmatrix} &= \left(HB \frac{n}{R} \frac{k_t}{R_a} \right)^{-1} G \begin{bmatrix} \dot{\overline{{}^m v_x}} \\ \dot{\overline{{}^m v_y}} \\ \dot{\overline{{}^m v_\phi}} \end{bmatrix} - \left(HB \frac{n}{R} \frac{k_t}{R_a} \right)^{-1} \begin{bmatrix} \overline{{}^m v_\phi}\,\overline{{}^m v_y} \\ -\overline{{}^m v_\phi}\,\overline{{}^m v_x} \\ 0 \end{bmatrix} \\
&\quad + B^T \frac{n}{R} \frac{R_a}{k_t} \left(D_m + \frac{k_t k_b}{R_a} \right) \begin{bmatrix} \overline{{}^m v_x} \\ \overline{{}^m v_y} \\ \overline{{}^m v_\phi} \end{bmatrix}
\end{aligned}
\tag{35}
$$

Additionally, the below quantities are mathematically defined

$$
\begin{bmatrix} e^m v_x \\ e^m v_y \\ e^m v_\phi \end{bmatrix} = \begin{bmatrix} {}^m v_x \\ {}^m v_y \\ {}^m v_\phi \end{bmatrix} - \begin{bmatrix} {}^m v_{xcom} \\ {}^m v_{ycom} \\ {}^m v_{\phi com} \end{bmatrix}
\tag{36}
$$

$$
\begin{bmatrix} \delta u_1 \\ \delta u_2 \\ \delta u_3 \end{bmatrix} = \begin{bmatrix} u_1 \\ u_2 \\ u_3 \end{bmatrix} - \begin{bmatrix} \overline{u_1} \\ \overline{u_2} \\ \overline{u_3} \end{bmatrix}
\tag{37}
$$

Linearizing along the nominal trajectories $\begin{bmatrix} \overline{{}^m v_x} & \overline{{}^m v_y} & \overline{{}^m v_\phi} \end{bmatrix}^T$ and $\begin{bmatrix} \overline{u_1} & \overline{u_2} & \overline{u_3} \end{bmatrix}^T$ based on the Eq. (19) in the similar procedure to outer loop control, so we attain the tracking error dynamics of inner loop

$$\begin{bmatrix} \dot{e}_{m_{v_x}} \\ \dot{e}_{m_{v_y}} \\ \dot{e}_{m_{v_\phi}} \end{bmatrix} = A_2 \begin{bmatrix} e_{m_{v_x}} \\ e_{m_{v_y}} \\ e_{m_{v_\phi}} \end{bmatrix} + B_2 \begin{bmatrix} \delta u_1 \\ \delta u_2 \\ \delta u_3 \end{bmatrix} \tag{38}$$

Where,

$$A_2 = G^{-1} \begin{bmatrix} 0 & \overline{m}_{v_\phi} & \overline{m}_{v_y} \\ -\overline{m}_{v_\phi} & 0 & \overline{m}_{v_x} \\ 0 & 0 & 0 \end{bmatrix} - G^{-1} HBB^T \frac{n^2}{R^2} \left(D_m + \frac{k_t k_b}{R_a} \right);$$
$$B_2 = G^{-1} HB \frac{k_t}{R_a} \frac{n}{R};$$

Similarly, we design a PI controller in order to stabilize the tracking error:

$$\begin{bmatrix} \delta u_1 \\ \delta u_2 \\ \delta u_3 \end{bmatrix} = -K_{P2} \begin{bmatrix} e_{m_{v_x}} \\ e_{m_{v_y}} \\ e_{m_{v_\phi}} \end{bmatrix} - K_{I2} \begin{bmatrix} \int e_{m_{v_x}}(t)dt \\ \int e_{m_{v_y}}(t)dt \\ \int e_{m_{v_\phi}}(t)dt \end{bmatrix} \tag{39}$$

From the Eq. (38) and (39), we get:

$$\dot{x}_i = A_{cli} x_i \tag{40}$$

Where,

$$x_i = \begin{bmatrix} \int e_{m_{v_x}}(t)dt \\ \int e_{m_{v_y}}(t)dt \\ \int e_{m_{v_\phi}}(t)dt \\ e_{m_{v_x}}(t) \\ e_{m_{v_y}}(t) \\ e_{m_{v_\phi}}(t) \end{bmatrix}; \quad \begin{array}{l} A_{cli} = \begin{bmatrix} O_3 & I_3 \\ -B_2 K_{I2} & A_2 - B_2 K_{P2} \end{bmatrix}; \\ O_3 \text{ is the } 3 \times 3 \text{ zero matrix;} \\ I_3 \text{ is the } 3 \times 3 \text{ identity matrix;} \end{array} \tag{41}$$

Design K_{P2} and K_{I2} to achieve the desired closed-loop tracking error dynamics

$$A_{cli} = \begin{bmatrix} O_3 & I_3 \\ diag[-a_{211} & -a_{221} & -a_{231}] & diag[-a_{212} & -a_{222} & -a_{232}] \end{bmatrix} \tag{42}$$

Where $a_{2j1} > 0$, $a_{2j2} > 0$, $j = 1, 2, 3$ are the coefficients of the desired closed-loop characteristic polynomial of each channel.

Based on the desired response and tracking error, we choose the suitable poles, and then define the coefficients of the desired closed-loop characteristic polynomial.

From Eq. (41) and Eq. (42), we get the gain K_{P2} and K_{I2} of the controller

$$K_{I2} = -B_2^{-1}diag[-a_{211} \quad -a_{221} \quad -a_{231}]$$
$$K_{P2} = B_2^{-1}(A_2 - diag[-a_{212} \quad -a_{222} \quad -a_{232}]) \tag{43}$$

Finally, the control input voltage to the motors is computed as

$$\begin{bmatrix} u_1 \\ u_2 \\ u_3 \end{bmatrix} = \begin{bmatrix} \delta u_1 \\ \delta u_2 \\ \delta u_3 \end{bmatrix} + \begin{bmatrix} \overline{u_1} \\ \overline{u_2} \\ \overline{u_3} \end{bmatrix} \tag{44}$$

4   Results of Study

In the simulation, we test the omni robot in two different trajectories: (i) a set-point trajectory, (ii) a circular trajectory. We can use the same control gains to follow these two trajectories.

The omni-directional robot has three dimensions of freedom. It can drive in translation and rotation independently. In the simulation, robot can be controlled to track translation and rotation trajectories at the same time.

4.1   Location Control

The initial position of robot is [0 0 0] and robot is move to stop at [0.05 0.05 π/4] from initial position. The simulation result is shown in Fig. 3.

4.2   Circular Trajectory

In the simulation, the mobile robot was controlled to accelerate from the initial position [0 0.5 0] and circularly move with the increasing angular speed from 0.05 rad/s to 1 rad/s. The center of the circular route was at [0 0] on Oxy coordinate and the robot body was maintained on the tangential direction of circle. The simulation outcomes are presented in Fig. 4 and Fig. 5. According to Fig. 4, the azimuth tracking error Ψ is displayed on the resolution of 0.001 radians, and the tracking error x and y in the resolution of 0.001 m.

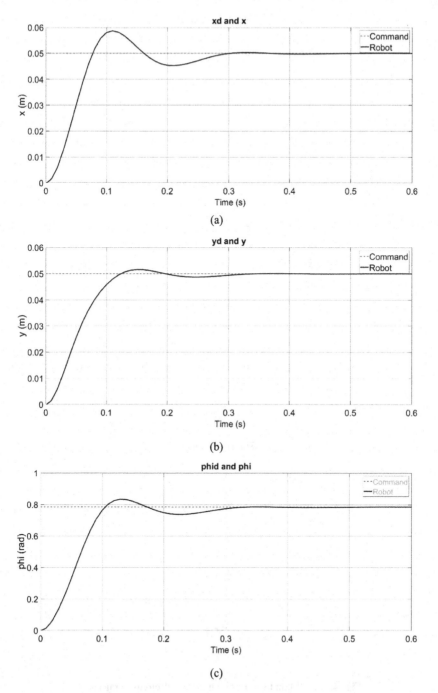

Fig. 3. Position control (a), (b) and (c) tracking result on x axis, y axis and rotational axis respectively

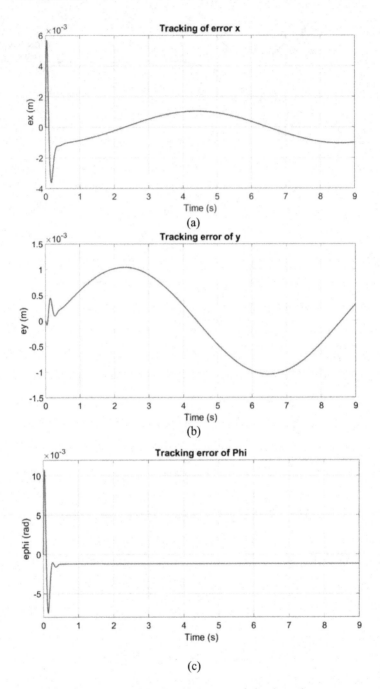

Fig. 4. (a), (b) and (c) Tracking errors of circular trajectory

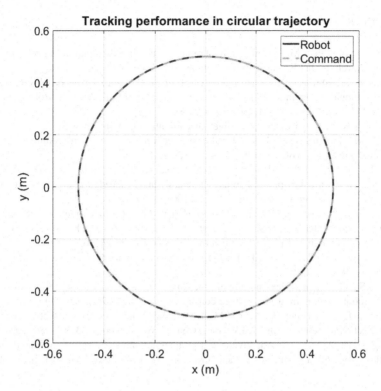

Fig. 5. Circular trajectory plot of omni robot

5 Conclusions

This article presented the equations of motion for an omni robot via dynamics, kinematics, and motor actuator to simulate the hardware platform. Based on these formulas, a nonlinear controller was implemented to design by TLC method. The simulation results showed that the robot can track trajectory with relative accuracy and the settling time is reasonable. Using this linearization approach, the controller gains could be tuned easily.

From the simulation results, it can be seen the proposed scheme for omni robot is robust, feasible and efficient. Future work is a must. Although our method is theoretically simulated, it is necessary to verify in a real hardware. Alternatively, a visual equipment with artificial intelligence or machine learning might be integrated into robot platform.

Acknowledgments. We acknowledge the support of time and facilities from Ho Chi Minh City University of Technology (HCMUT), VNU-HCM for this study.

References

1. Pin, F.G., Killough, S.M.: A new family of omnidirectional and holonomic wheeled platforms for mobile robots. IEEE Trans. Robot. Autom. **10**(4), 480–489 (1994)
2. Watanabe, K., Shiraishi, Y., Tzafestas, S.G., Tang, J., Fukuda, T.: Feedback control of an omnidirectional autonomous platform for mobile service robots. J. Intell. Rob. Syst. **22**(3), 315–330 (1998). https://doi.org/10.1023/A:1008048307352
3. Watanabe, K.: Control of an omnidirectional mobile robot. In: 1998 Second International Conference on Knowledge-Based Intelligent Electronic Systems. Proceedings KES 1998 (Cat. No. 98EX111), vol. 1, pp. 51–60, April 1998
4. Zhu, J.J., Banker, B.D., Hall, C.E.: X-33 ascent flight controller design by trajectory linearization-a singular perturbational approach. In: Proceedings of AIAA Guidance, Navigation, and Control Conference, August 2000
5. Nguyen, T.P., Nguyen, H., Phan, V.H., Ngo, H.Q.T.: Modeling and practical implementation of motion controller for stable movement in a robotic solar panel dust-removal system. Energy Sources Part A Recovery Utilization Environ. Eff. 1–22 (2021)
6. Zheng, Z.W., Huo, W.: Trajectory tracking control for a stratospheric airship. Control Decis. **26**(10), 1479–1484 (2011)
7. Choi, J.S., Kim, B.K.: Near minimum-time direct voltage control algorithms for wheeled mobile robots with current and voltage constraints. Robotica **19**(1), 29 (2001)
8. Huang, J.-T., Van Hung, T., Tseng, M.-L.: Smooth switching robust adaptive control for omnidirectional mobile robots. IEEE Trans. Control Syst. Technol. **23**(5), 1986–1993 (2015)
9. Huang, H.-C., Tsai, C.-C.: Adaptive trajectory tracking and stabilization for omnidirectional mobile robot with dynamic effect and uncertainties. IFAC Proc. Vol. **41**(2), 5383–5388 (2008)

Adaptive Deep Learning Technique to Predict Student's Graduation Results

Nguyen Quoc Viet[1], Vo Pham Tri Thien[2],
and Nguyen Thanh Binh[3,4(✉)]

[1] Student Affairs Office, Ho Chi Minh City University of Education,
Ho Chi Minh City, Vietnam
vietnqu@hcmue.edu.vn
[2] Undergraduate Studies Office, Ho Chi Minh City University of Education,
Ho Chi Minh City, Vietnam
thienvpt@hcmue.edu.vn
[3] Department of Information Systems, Faculty of Computer Science
and Engineering, Ho Chi Minh City University of Technology (HCMUT),
268 Ly Thuong Kiet Street, District 10, Ho Chi Minh City, Vietnam
ntbinh@hcmut.edu.vn
[4] Vietnam National University Ho Chi Minh City, Linh Trung Ward,
Thu Duc District, Ho Chi Minh City, Vietnam

Abstract. Analyzing educational data techniques helps educational institutions predict the final graduation results of students based on the average score in the first semesters of the course. To predict student's graduation results, this research proposed the method to predict includes five steps: data collection, data preprocessing, predict student's graduation results model, evaluating model and deployment model. The proposed method is based on the average academic points data of the six main semesters out of a total of eight main semesters of students to predict their graduation result with the accuracy of about 90.53%. To objectively evaluate the effectiveness of the proposed method, the results of the proposed method, which compared with the other methods, are better than the other methods.

Keywords: Data analysis · Student's graduation result · Deep learning · Data mining

1 Introduction

Higher education is one of the most important forms of education in the country. Students will have tremendous opportunities of seeking jobs suitable to their specialized majors after the completion of the university education programs. In our country, according to statistics of the Ministry of Education and Training, there are approximately one million students enrolled by about 237 education institutions every year [1]. Many newly enrolled students will be a big concern for the training quality to which the educational institutions should pay attention. Therefore, higher education institutions need to forecast the learning situation of students to timely support students with weak academic ability to keep up with their learning progress.

© ICST Institute for Computer Sciences, Social Informatics and Telecommunications Engineering 2021
Published by Springer Nature Switzerland AG 2021. All Rights Reserved
P. Cong Vinh and N. Huu Nhan (Eds.): ICTCC 2021, LNICST 408, pp. 67–75, 2021.
https://doi.org/10.1007/978-3-030-92942-8_6

Today, with the advent of machine learning techniques, it is possible for users to use them to mine large data sets to find out their impact on the area of life. Educational data mining is also an area that many scientists have been researching. Using educational data in universities and colleges to analyze, predict, and support decision making has become one of the most important contributions to improve the quality of education in universities and colleges. One of the essential factors that directly affect the graduation results of students is the average score of the courses. For students with a good result, a high-grade point average (GPA) reflects their entire study procedure, and they easily achieve the desired graduation results. For students with average, good, or likely failure to graduate, we can use GPA data to give them predictions about graduation or future outcome within the following school year to change their study methods to get the results they want and graduate higher classification.

With such requirements, applying data mining techniques, especially Educational data mining to predict the students' graduation result will help the schools figuring out the opportunities as well as the risks to improve not only teaching quality but also the studying condition at the university. This research proposed a method to predict student's graduation result. The proposed model includes five steps: data collection, data preprocessing, predict student's graduation results model, evaluating model and deployment model. We use the data of the average academic points from the six main semesters out of a total of eight main semesters of students to predict their graduation result. Accuracy criteria proposed to evaluate the results between our proposed method and the recent other methods. The rest of the paper is organized as the following: the literature review is presented in Sect. 2. The proposed method for predicting outcomes for students about to graduate in 2021 is presented in Sect. 3. The result analysis is presented in Sect. 4. Finally, Sect. 5 is conclusions and future works.

2   Literature Review

Educational data mining (EDM) is a field of data mining, with superabundant data from the education industry, schools can apply data mining techniques to analyze and predict students' achievement [2–4]. Early prediction of student results can help administrators make the necessary decisions at the right time and create the right training plan to improve students' academic success rates.

Many researchers used many methods to predict how successful students will be at the time of receiving their graduation certificate or in the early years of the course [2, 5, 6]. According to previous relevant research, students' academic success is defined by factors such as academic achievement, demographics, satisfaction, competencies, acquisition of skills, attainment of learning objectives during their academic careers. Besides, the critical factors affecting students' academic success is the grade point average of the course [7], prior academic achievement and student demographics [2], which presented 69% of the total factors affecting student learning out-comes, the remaining 31% are related factors such as e-learning activities, psychological attributes, and environments.

There are many approaches to predict student learning outcomes, in which, the classification model and regression model are the most used [8–10]. The algorithms used in the classification model are decision tree, Bayesian, artificial neural networks, rule learner's, ensemble learning, K-Nearest neighbor. These algorithms account for 95% of prior studies [2]. For forecasting student results, recent studies have used techniques such as Decision trees [12, 14, 18], Deep learning, Neural network, Naïve Bayes, Support vector machine [15]. In addition, some other models are also used for student results prediction like Generalized linear model [16]. Data mining techniques are used to predict student's performance in the first or second year in the university [17].

Recently, machine learning techniques are used to analyze data and predict students' results [3, 11–13]. Among research about academic success prediction, GPA is one of the prerequisite factors affecting the academic achievement of students [2, 14]. Deep learning is also a widely and used technique and remains an important role in predicting student final GPA [19]. However, each method has its own advantages and disadvantages.

3 Methodology

To carry out this study, this paper proposes a model to predict outcomes for students about to graduate in 2021. The proposed model includes five steps: data collection, data preprocessing, predict student's graduation results model, evaluating model and deployment model which is presented as Fig. 1. The details of the steps in the proposed model are presented below.

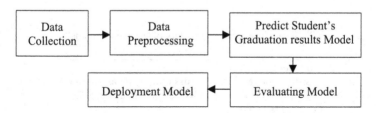

Fig. 1. The steps in the proposed model

(i) Data Collection: The data provided by our university, which is blur, includes data of graduates in 2018, 2019, and 2020 for training. Besides that, data of students, who are to graduate in 2021, are used to deploy the trained model. Our university has approximately 2.000 graduates every year. Therefore, predicting graduation results will significantly help students while studying at the university. In this research, we use the data of the average academic points from the 6 main semesters of students provided by the university. Our data has 6064 students with their data property as: Student ID, Gender, Student's GPA in semester 1, 2, 3, 4, 5, 6 and final graduation result.

(ii) Data Preprocessing: The provided data is stored in many individual datasets. It is heterogeneous in structure and properties. So, we have processed the data by selecting the necessary attributes for this study and saving them to a single dataset so that training models can be deployed.

(iii) Predict student's graduation results model: According to the related work which presented in the previous section, the neural network method has many advantages in predicting student's graduation. Therefore, we choose the method to solve this problem using neural networks.

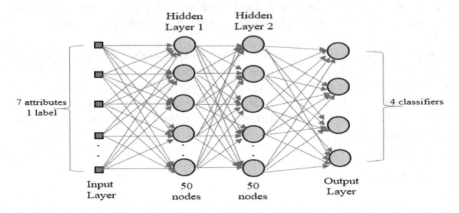

Fig. 2. Predict student's graduation results model

Based on the dataset to be evaluated, the goal of the out and the requirements of prediction of the student's graduation, the neural network model selected and presented as Fig. 2.

Table 1. The other parameters of the predict model.

* *The parameters:*
+ Loss function: Multiclass Cross Entropy (Classification)
+ Optimization method: Stochastic gradient descent
+ Learning rate: 0,005
+ Number of hidden layers: 2
+ Neuron in layers: 50
+ Epochs = 300
+ Epsilon = 1.0e-6
+ Beta1 = 0,9
+ Beta2 = 0,999
* *Parameter for Hidden Layer 1 and Hidden Layer 2:*
+ Nodes = 50
+ Activation Function: Rectified Linear Unit
* *Output Layer:*
+ Nodes = 4
+ Activation Function: Softmax

As presented as Fig. 2, the input layer includes seven attributes as: Gender, Student's GPA in semester 1, 2, 3, 4, 5, 6 and final graduation result (one label). With the hidden layer 1 and 2, every layer has 50 nodes. The output layer is divided into four classifiers: Average, Good, Very Good, Excellent. Because the students do not graduate if their GPA is below average. The other parameters of the predict model selected as Table 1.

(iv) Evaluating Model: We will evaluate the performance of the model and against the other models such as: decision tree, random forest. This part will present detail in Sect. 4.

(v) Deployment Model: After conducting experiments with data of graduates in 2018, 2019, and 2020, we will deploy the above models for predicting graduation grades for graduates in 2021.

In the next part, we implement the proposed method and results analysis.

4   Result Analysis

4.1   Material

Our experiments are developed in Python, with the computer of Intel core i5 (4 CPUs), 2.5 GHz CPU, 4 GB DDR2 memory. We collected the data which was presented in Sect. 3 (data collection and data preprocessing steps) for our experimentation. Data collection consists of two datasets: dataset A and dataset B. Detail of these dataset include:

+ Dataset A includes 4648 rows of data equivalent to 4648 students who graduated in 2018, graduated in 2019 and graduated in 2020. Every data in dataset A has nine attributes, including: Student ID, Gender, Student's GPA in semester 1, 2, 3, 4, 5, 6 and final graduation result.

+ Dataset B includes 1416 rows of data equivalent to 1416 students will graduate in 2021. Every data in dataset B has eight attributes, including: Student ID, Gender, Student's GPA in semester 1, 2, 3, 4, 5 and 6.

Based on the rules of the university, graduate classification of students depends on their GPA. Specifically, we categorize the result into four classifications as:

+ The first classification with GPA between 2.0 and 2.49 is Average.

+ The second classification with GPA between 2.5 and 3.19 is Good.

+ The third classification with GPA between 3.2 and 3.59 is Very Good.

+ The last classification with GPA between 3.6 and 4.0 is Excellent.

4.2   Experimentation and Evaluation Results

Firstly, we evaluated to predict student's graduation results model on dataset A. This dataset is divided into 60% for the training dataset and 40% for the test dataset (equivalent to 1859 students). After conducting training and testing the models, the accuracy metric proposed to evaluate results of the prediction model. With some training epochs and learning rate, the model will be fit for 300 training epochs and learning rate is 0,005. The confusion matrix of the proposed method is shown in Table 2. We see that the proposed method has good results.

Table 2. Confusion matrix for the proposed method in dataset A

Prediction level	True good	True very good	True excellent	True average
Average	1	0	0	6
Good	1092	70	0	13
Very good	62	540	26	0
Excellent	0	4	45	0

To objectively evaluate the effectiveness of the proposed method, we compared the results of the proposed method with Decision Tree method [8], Random Forest method [20] by the evaluation as accuracy criteria. Table 3 and Table 4 presented the confusion matrix of the Decision Tree method and Random Forest method with the above dataset. The accuracy of the proposed method with the other methods presented as Table 5.

Table 3. Confusion matrix for the Decision Tree method in dataset A

Prediction level	True good	True very good	True excellent	True average
Average	12	0	0	13
Good	1021	98	0	6
Very good	121	476	18	0
Excellent	1	40	53	0

Table 4. Confusion matrix for the Random Forest method in dataset A

Prediction level	True good	True very good	True excellent	True average
Average	13	1	0	8
Good	984	87	0	11
Very good	158	498	18	0
Excellent	0	28	53	0

In Table 5, the accuracy of the proposed method is 90.53% while the accuracy of the Decision Tree method and Random Forest method are 84.08% and 83%, respectively. So, the results of the proposed method are better than the other methods.

Table 5. Accuracy of the proposed method with the other methods in dataset A

Evaluation criteria	Decision tree method	Random forest method	Proposed method
Correct classified	1563	1543	1683
Wrong classified	296	316	176
Accuracy	84,08%	83,00%	90,53%
Error	15,92%	17,00%	9,47%

Secondly, we evaluated to predict student's graduation results model on dataset B. The experimentation is implemented to predict the learning outcomes for 1416 students who are preparing to graduate in 2021. The dataset B has a structure and properties quite like the dataset A. The difference in this dataset is that there is no attribute of the students' final graduation results.

With the prediction for dataset B, the results in Table 6 show that the predictive model gives relatively good results and can be applied to problems using average scores to predict graduation outcomes for students.

Table 6. Number of true and false prediction in dataset B

Prediction level	Decision tree	Random forest	Proposed method
True prediction	1194	1171	1275
False prediction	222	245	141

Table 7. Accuracy of the proposed method with the other methods in dataset B

Evaluation criteria	Decision tree	Random forest	Proposed method
Accuracy	84,32%	82,7%	90,04%
Error	15,68%	17,3%	9,96%

Table 7 presented the results of predicting our proposed method with other methods. In the dataset B with fewer data rows than the dataset A, the accuracy among the models is not significantly different. The highest accuracy belongs to the proposed method at a rate of 90.04% while the accuracy of the Decision Tree method and Random Forest method are 84.32% and 82.7%, respectively. From the experimentation on dataset A and dataset B, the results of the proposed method are better than the other methods.

5   Conclusions and Future Works

Analyzing educational data techniques helps educational institutions predict the final graduation results of students based on the average score in the first semesters of the course. The experiment shows that using the average scores in the first six semesters of the course will help educational institutions to categorize students. From that, they can intervene in the students' study procedure, encourage students to improve their study results and to desired academic success. The proposed method used in this study gives pretty good output in terms of performance. Along with training and testing the models, the research has also deployed models to predict the grade results for students graduating in 2021 with the high prediction accuracy. In the future work, besides using the average score and the grade classification, we need to use more data from the results of each subject to increase the results predicted.

Acknowledgment. We would like to thank HCMC University of Education for the data in this research, so this research is able to be done and hope can bring benefit for all their students. We acknowledge the support of time and facilities from Ho Chi Minh City University of Technology (HCMUT), VNU-HCM for this study.

References

1. Ministry of Education and Training. https://moet.gov.vn/thong-ke/Pages/thong-ko-giao-duc-dai-hoc.aspx?ItemID=7389. Accessed 4 July 2021
2. Alyahyan, E., Düştegör, D.: Predicting academic success in higher education: literature review and best practices. Int. J. Educ. Technol. High. Educ. **17**, 1–21 (2020)
3. Romero, C., Ventura, S.: Data mining in education. Wiley Interdiscipl. Rev. Data Min. Knowl. Discov. **3**(1), 11–27 (2013)
4. Martins, M.P.G., Miguéis, V.L., Fonseca, D.S.B., Alves, A.: A Data mining approach for predicting academic success – a case study. In: Rocha, Á., Ferrás, C., Paredes, M. (eds.) ICITS 2019. AISC, vol. 918, pp. 45–56. Springer, Cham (2019). https://doi.org/10.1007/978-3-030-11890-7_5
5. Acharya, A., Sinha, D.: Early prediction of students performance using machine learning techniques. Int. J. Comput. Appl. **107**(1), 37–43 (2014). https://doi.org/10.5120/18717-9939
6. Tanuar, E., Heryadi, Y., Abbas, B.S., Gaol, F.L.: Using machine learning techniques to earlier predict student's performance. In: 2018 Indonesian Association for Pattern Recognition International Conference (INAPR), pp. 85–89. IEEE (2018)
7. York, T.T., Gibson, C., Rankin, S.: Defining and measuring academic success. Pract. Assess. Res. Eval. **20**, 5 (2015). https://doi.org/10.7275/hz5x-tx03
8. Adekitan, A.I., Odunayo, S.: The impact of engineering students' performance in the first three years on their graduation result using educational data mining. Heliyon **5**(2), e01250 (2019). https://doi.org/10.1012/j.heliyon.2019.e01250
9. Almarabeh, H.: Analysis of students' performance by using different data mining classifiers. Int. J. Mod. Educ. Comput. Sci. **9**, 9–15 (2017). https://doi.org/10.5815/ijmecs.2017.08.02
10. Asif, R., Merceron, A., Ali, S.A., Haider, N.G.: Analyzing undergraduate students' performance using educational data mining. Comput. Educ. **113**, 177–194 (2017)
11. Garg, R.: Predicting student performance of different regions of Punjab using classification techniques. Int. J. Adv. Res. Comput. Sci. **9**, 236–241 (2018)
12. Hamoud, A., Hashim, A.S., Awadh. W.A.: Predicting student performance in higher education institutions using decision tree analysis. Int. J. Interact. Multim. Artif. Intell. **5**, 26–31 (2018)
13. Xu, J., Moon, K.H., Van Der Schaar, M.: A machine learning approach for tracking and predicting student performance in degree programs. IEEE J. Select. Top. Sig. Process. 11, 742–753 (2017)
14. Al-Barrak, M.A., Al-Razgan, M.: Predicting students final GPA using decision trees: a case study. Int. J. Inf. Educ. Technol. **6**(7), 528–533 (2016). https://doi.org/10.7763/IJIET.2016.V6.745
15. Babić, I.Đ.: Machine learning methods in predicting the student academic motivation. Croat. Oper. Res. Rev. **8**, 443–461 (2017)
16. Kumar, A., Naughton, J., Patel, J.M.: Learning generalized linear models over normalized data. In: Proceedings of the 2015 ACM SIGMOD International Conference on Management of Data (2015). https://doi.org/10.1145/2723372.2723713

17. Adekitan, A.I., Noma-Osaghae, E.: Data mining approach to predicting the performance of first year student in a university using the admission requirements. Educ. Inf. Technol. **24**(2), 1527–1543 (2018). https://doi.org/10.1007/s10639-018-9839-7
18. Pandey, M., Sharma, V.K.: A decision tree algorithm pertaining to the student performance analysis and prediction. Int. J. Comput. Appl. **61**(13) (2013)
19. Sultana, J., Usha, M., Farquad, M.A.H.: Student's performance prediction using deep learning and data mining methods. Int. J. Recent Technol. Eng. **8**(1S4) (2019). ISSN: 2277-3878
20. Jayaprakash, S., Krishnan, S., Jaiganesh, V.: Predicting students academic performance using an improved random forest classifier. 2020 International Conference on Emerging Smart Computing and Informatics (ESCI), pp. 238–243. IEEE (2020)

An Approach for Multiple Choice Question Answering System

Dinh-Huy Vo[1], Anh-Khoa Do-Vo[1], Tram-Anh Nguyen-Thi[2], and Huu-Thanh Duong[1]([✉])

[1] Faculty of Information Technology, Ho Chi Minh City Open University, Ho Chi Minh City, Vietnam
{1851010052huy,1851010057khoa,thanh.dh}@ou.edu.vn
[2] Department of Fundamental Studies,
Ho Chi Minh City Open University, Ho Chi Minh City, Vietnam
tramanh.nguyen@ou.edu.vn

Abstract. Multiple choice question (MCQ) system, which is a form of question answering system, includes a question, a set of choices and the correct answers from these choices. The rapidly growth of linguistics models improves natural language understanding and motivates the automatic systems in natural language processing. This paper proposes an approach based on the Language Model for MCQ system as incomplete questions in English tests. This not only reduces the burden of human experts to teach their students, but also is useful for self-studying the students. We perform many experiments to evaluate the effectiveness and achieve 85.45% accuracy for our proposal.

Keywords: MCQ · BERT · MLM · Multiple choice · Language model · Unigram · Bigram · Masked Language Model

1 Introduction

Question answering (QA), which is an essential task in natural language processing (NLP), gives the succinct and short answers for the user's query in natural language, MCQ system is a subdomain of QA system where a question and the limited multiple choices are given and we need to choose the correct answers among these choices in particular context without user intervention.

MCQ plays a major role in educational assessment. With abundant resources and rapid growth of the pre-trained models in English, the machine totally answers automatically the multiple choice questions in the specific context. Using the automatic system not only reduces the burden of human experts to teach and evaluate their students, but also is useful for self-studying. In this paper, we develop the MCQ system for the forms of incomplete questions in English exams in any subjects such as grammar, nouns, pronouns, vocabularies, verb tenses based on our training corpus which is collected from the well-known websites.

© ICST Institute for Computer Sciences, Social Informatics and Telecommunications Engineering 2021
Published by Springer Nature Switzerland AG 2021. All Rights Reserved
P. Cong Vinh and N. Huu Nhan (Eds.): ICTCC 2021, LNICST 408, pp. 76–82, 2021.
https://doi.org/10.1007/978-3-030-92942-8_7

Our main contribution has performed the existing methods for multiple choice question problems such as Language Model (unigram, bigram) and Masked Language Model (MLM), proposed and evaluated an approach as an ensemble of above methods to improve the accuracies of multiple choice question problems.

The rest of this paper is organized as follows. Section 2 presents the related works. Section 3 presents methods that are used for the automated English multiple-choice test system. Next, Sect. 4 shows the experimental results. Finally, the conclusion and our future direction of system development will be presented in Sect. 5.

2 Related Works

QA systems have been concerned and has many proposals for it such as Duong and Ho [1] used Language Model for QA system in legal documents, Duong and Hoang [2] combined of machine learning classifiers in news classification, Y. Sharma et al. [3] applied the deep learning approaches for QA system, W. Yu et al. [4] used transfer learning with ALBERT to build QA system in technical domains.

As a form of QA system, MCQ has emerged and attracted many research teams to resolve the multiple choice questions in recent years. A. Chaturvedi et al. [5] proposed to use CNN model to answer the multiple choice questions in particular domains. The accuracies are better than the baseline results as LSTM model in two datasets as Textbook question answering (TQA) and SciQ datasets. K. Moholkar et al. [6] proposed an ensemble of models approach, including LSTM, hybrid LSTM-CNN and multiplayer perceptron models to predict the correct answer. LSTM and hybrid LSTM-CNN models are firstly trained parallelly, then the multilayer perceptron model predicts the correct answer. This approach obtains higher accuracies than using the models separately. R. Chitta and A.K. Hudek [7] developed the question answering system for multiple choice questions in legal contracts. They have extracted the relevance text in contracts and used a multi-class classifier to choose the correct answer based on the extracted text.

Language Model (LM) is a statistical approach learning the probability distribution over a sequence of words based on the existing corpus. This model is widely used in information retrieval. This refers to an effective solution to predict the correct choice of the multiple choice question in incomplete form based on its probability distribution.

J. Devlin et al. at Google API Language published BERT (Bidirectional Encoder Representations from Transformers) in [8], this applies the bidirectional training of a transformer instead of only looking at the single directional training like the previous efforts. In this paper, they also introduced a novel technique named Masked Language Model (MLM). MLM is an important model of BERT, here we give a sentence to which some words are attached [MASK] and then predict these words. The benefit of BERT is that it predicts the hidden word through the context of words on the right and left so when we cover a word, it can predict the hidden word effectively because based on the context of the whole question. Thus, MLM is a suitable method to resolve multiple choice questions in incomplete form.

3   The Approach

We use a statistical approach including LM (unigram and bigram) and MLM. These models are conducted separately to evaluate the effectiveness. Then we propose an ensemble of two models to improve the performances of MCQ problems. The experiments present this approach achieves better accuracy significantly than applying the models separately.

3.1   Language Model

The main purpose of LM is to calculate probability distribution over a sequence of words: w_1, w_2, w_3, ...w_n based on the Naïve Bayes formula:

$$P(w_1 w_2...w_n) = P(w_1) \times P(w_2|w_1) \times ... \times P(w_n|w_1 w_2...w_{n-1}) \qquad (1)$$

According to this formula, LM needs to use a large amount of memory to store the results of the string. This would be impossible if the length of the string is a paragraph or even more because the length of a natural text is infinite. The n-gram models (or n-level Markov Models) will solve this problem by calculating the probability of the sequence with lower memory. The occurrence probability of (w_m) will depend on n consecutive words preceding it, not on the entire sequence. Thus we have the probability of word (w_m) calculated as follows:

$$P(w_1 w_2...w_m) = P(w_1) \times P(w_2|w_1) \times ... \times P(w_m|w_{m-n-1} w_{m-n}...w_{m-1}) \quad (2)$$

For $n = 1$ calls unigram which is the simplest model in n-gram. It evaluates the probability of each word in a sentence independently as below formula.

$$P(w_i) = \frac{C(w_i)}{N} \qquad (3)$$

Where $P(w_i)$ is the probability of the word w_i, $C(w_i)$ is the number of occurrences of w_i, N is the total number of words in the training corpus.

For $n = 2$ calls bigram, this will consider the probability of occurrence of two consecutive words in a sentence according to the formula:

$$P(w_i|w_{i-1}) = \frac{C(w_{i-1} w_i)}{C(w_{i-1})} \qquad (4)$$

3.2   Masked Language Model

Masked Language Model is the most important model structure of BERT. MLM combines Transformer encoder and masked tokens can predict a missing token in a sentence. The output of MLM is the word embeddings of corresponding tokens which feed to a simple softmax classifier to get the final prediction.

The missing token is represented by the [MASK] symbol in a sentence and MLM predicts the suitable tokens for replacement. BERT can predict through the context of words in its right and left directions. It assumes the missing token is at position t (w_t) , the context tokens to predict includes w_1, w_2, ..., w_{t-1}, $w_t + 1$, ..., w_n (n is the number of tokens in a sentence). This considers the context of the whole sentence, so this can predict the missing token.

3.3   Our Proposal

MLM uses bidirectional context words of the hidden word and considers the context of the whole sentence, so this is a good approach for incomplete forms. However, for multiple choice questions, the correct answer only limits specific choices. As a result, MLM cannot sometimes give the final answer since the predicted words do not match any choices. Otherwise, n-gram models always obtain the final answer, but they only consider the previous words of the hidden word. Thus, our proposal gives the combination of n-gram (unigram and bigram) models and MLM. In order to determine the correct answer, we use a formula based on a linear equation:

$$f(x,y) = \lambda x + (1 - \lambda)y \tag{5}$$

Where λ is a coefficient between [0–1], x is the score of choosing the answer by MLM, y is the score of choosing the answer of the unigram or bigram.

4   The Experiments

For the training corpus, we have extracted 207,319 documents (magazines, articles, etc.) from the well-known websites in English and prepared 502 multiple choice questions with answers for testing to evaluate our approach (Table 1 shows some samples of testing data). The experiments are firstly conducted on unigram, bigram, MLM separately, the selected answer has the highest score from these models. Afterward, we combine these models, including (unigram and MLM) and (bigram and MLM), the scores are determined by formula (5).

Table 2 presents the accuracies of various experiments, where unigram and bigram always give the answer and obtain 29.48% and 57.37% accuracies respectively. The multiple choice questions demand the formal structures and grammars, so the order of words is essential in text. The unigram does not concern this, leading to this model obtaining a low accuracy in this problem. The bigram model calculates the probability of a word based on the context of its previous word, it improves accuracy significantly compared to the unigram model.

For MLM, the selected answer is suggested by MLM and matches one of the choices of the question, its accuracy is 44.02%. This model predicts the hidden word based on its bidirectional words. However, there are many cases in which it is impossible to produce the results since some suggested words of MLM not matching any choices of the question.

The combination of unigram and MLM obtains 82.86% accuracy with λ in (0.2, 0.3, 0.4, 0.5). The highest result, which is the combination of bigram and MLM, is 85.45 % accuracy with λ as 0.99. In order to select a good λ value, we perform many experiments with various λ values, the Fig. 1 and Fig. 2 show the results.

Table 1. The samples of testing data.

Question	A	B	C	D	Answer
The simplest way to reduce your ... footprint is to cycle to school	Carbon	Chemical	Chemistry	Dioxide	Carbon
My brother has to work ... a night shift once a week	In	On	At	By	On
It is hardly possible to ... the right decision all the time	Do	Arrive	Make	Take	Make
Buying organic food is better for the environment because it uses less ...	Fertilizer	Fertilize	Fertilizes	Fertilized	Fertilizer
No one on the plane was alive in the accident last night, ... they?	Wasn't	Weren't	Were	Was	Were
The movie is ... on a true story	Based	Keen	Hang	Let	Based
My father learned to play ... piano when he was five years old	A	An	The	Was	The
The dog was frightened by the sound of the thunder	Belt	Bolt	Bell	Bull	Bolt
It turned out that i ... have bought frank a present after all	Oughtn't	Mustn't	Needn't	Mightn't	Needn't

Table 2. The accuracies of various experiments.

Unigram	Bigram	MLM	Unigram + MLM ($\lambda = 0.50$)	Bigram + MLM ($\lambda = 0.99$)
29.48%	57.37%	44.02%	82.86%	*85.45%*

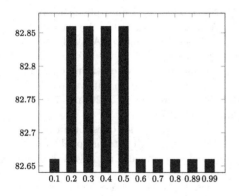

Fig. 1. Unigram combined with MLM with various λ values.

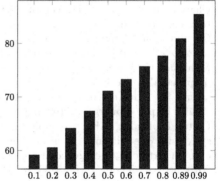

Fig. 2. Bigram combined with MLM with various λ values.

5 Conclusions and Future Works

In this paper, we have built an effective MCQ system for multiple choice questions in incomplete form for the English test. The experimental section conducts several experiments based on LM (unigram, bigram) and MLM. Moreover, we combine the LM and MLM and achieve promising accuracy.

In future works, we will broaden the system with more forms of multiple choice questions, resolve multiple choice questions of reading comprehension form based on the context of a defined paragraph or answer many multiple choice questions depending on an incomplete paragraph (such as incomplete part of the TOEIC test).

References

1. Duong, H.-T., Ho, B.-Q.: A Vietnamese question answering system in Vietnam's legal documents. In: Saeed, K., Snášel, V. (eds.) CISIM 2014. LNCS, vol. 8838, pp. 186–197. Springer, Heidelberg (2014). https://doi.org/10.1007/978-3-662-45237-0_19

2. Duong, H., Truong Hoang, V.: Question answering based on ensemble classifier for university enrolment advising. In: 2019 11th International Conference on Knowledge and Smart Technology (KST), Phuket, Thailand, 2019, pp. 35–39 (2019). https://doi.org/10.1109/KST.2019.8687616

3. Sharma, Y., Gupta, S.: Deep learning approaches for question answering system. Procedia Comput. Sci. **132**, 785–794 (2018). https://doi.org/10.1016/j.procs.2018.05.090

4. Yu, V., et al.: A technical question answering system with transfer learning. In: Proceedings of the 2020 Conference on Empirical Methods in Natural Language Processing: System Demonstrations, pp. 92–99 (2020) https://doi.org/10.18653/v1/2020.emnlp-demos.13

5. Chaturvedi, A., Pandit, D., Garain, U.: CNN for text-based multiple choice question answering. In: Proceedings of the 56th Annual Meeting of the Association for Computational Linguistics (Volume 2: Short Papers, Melbourne, Australia, 2018, pp. 272–277 (2018). https://doi.org/10.18653/v1/P18-2044
6. Moholkar, K., Patil, S.H.: Multiple choice question answer system using ensemble deep neural network. In: 2020 2nd International Conference on Innovative Mechanisms for Industry Applications (ICIMIA), Bangalore, India, 2020, pp. 762–766 (2020). https://doi.org/10.1109/ICIMIA48430.2020.9074855
7. Chitta, R., Hudek, A.K.: A reliable and accurate multiple choice question answering system for due diligence. In: Proceedings of the Seventeenth International Conference on Artificial Intelligence and Law, Montreal QC Canada, pp. 184–188, June 2019. https://doi.org/10.1145/3322640.3326711
8. Devlin, J., Chang, M.-W., Lee, K., Toutanova, K.: BERT: pre-training of deep bidirectional transformers for language understanding, May 2019. arXiv:1810.04805 [cs]. Accessed 01 May 2021, arXiv:1810.04805
9. Salazar, J., Liang, D., Nguyen, T.Q., Kirchhoff, K.: Masked language model scoring. In: Proceedings of the 58th Annual Meeting of the Association for Computational Linguistics, pp. 2699–2712 (2020). https://doi.org/10.18653/v1/2020.acl-main.240

Forecast of the VN30 Index by Day Using a Variable Dimension Reduction Method Based on Kernel Tricks

Thanh Do Van[1(✉)] and Hai Nguyen Minh[2]

[1] Faculty of Information Technology, Nguyen Tat Thanh University,
Ho Chi Minh City, Vietnam
dvthanh@ntt.edu.vn
[2] Faculty of Basic Sciences, Industrial University of Ho Chi Minh City,
Ho Chi Minh City, Vietnam
nguyenminhhaidhcn@iuh.edu.vn

Abstract. The stock market is an important channel in attracting investment capital. Its stock index and the stock prices of some large-cap enterprises are often leading indicators of the economy. Stock index forecasting is always a matter of interest to researchers and applications. There is now much research on stock market forecasting using technical or fundamental analysis methods through various soft computing techniques or algorithms.

In Vietnam, the VN30 stock index is calculated from the trading results of the 30 largest capitalization enterprises of the economy. The capitalization rate of these 30 enterprises accounts for more than 80% of the capitalization rate of the stock market. Investors and economic analysts often use this index's forecast information when making investment decisions or to analyze and forecast the activities of the economy. However, the information and data affecting the fluctuation of the VN30 stock index are often very large, which is the main obstacle when forecasting this index.

The purpose of this article is to perform a daily forecast of the VN30 index on a large dataset of predictors by using econometric techniques on factors extracted from the dataset of the predictors using the variable dimension reduction method based on kernel tricks. The forecast accuracy according to such an approach is very high.

Keywords: Time series · Dimensionality reduction · Kernel tricks · PCA · Stock index forecast

1 Introduction

In the article [1], the authors made a systematic and critical review of about one hundred and twenty-two (122) studies reported in academic journals, including journal articles, conference proceedings articles, doctoral dissertations, or supplementary unpublished academic working articles and reports between 2007 and 2018 in the field of the stock market forecast using machine learning. This article grouped these studies into three categories, namely technical, fundamental, and hybrid analysis, and showed that 66% of

© ICST Institute for Computer Sciences, Social Informatics and Telecommunications Engineering 2021
Published by Springer Nature Switzerland AG 2021. All Rights Reserved
P. Cong Vinh and N. Huu Nhan (Eds.): ICTCC 2021, LNICST 408, pp. 83–94, 2021.
https://doi.org/10.1007/978-3-030-92942-8_8

considered documents are based on technical analysis, whip 23% and 11% on the fundamental analysis and hybrid analysis, respectively. Support vector machines and neural networks are the most used machine learning algorithm for stock market forecasts.

Technical analysis is the method to predict the stock market by learning charts that portray the historical market prices and technical indicators. Some of the technical indicators used in technical analysis are discussed in [2–4]. The fundamental analysis uses the economic standing of the enterprise, the board of directors, financial status, firm's yearly report, balance sheets, income reports, disasters, and political data to predict future stock price [2, 5–7]. Technical analysis uses structured data (quantitative data), especially scalar time series data, while fundamental analysis uses unstructured data (qualitative data). Fundamental analysis has been developed quite strongly in recent years thanks to many machine learning algorithms, especially text mining techniques [8] that can automatically make predictions with pretty high accuracy on unstructured data. However, fundamental analysis is helpful for long-term stock price movements but not short-term stock price changes [9]. In the hybrid analysis, both quantitative and qualitative data are used to predict the stock market. The manifestation of this combination is that the data used to forecast the stock market need to include at least 2 data types of the following data types: past stock data, social media, financial news, and macroeconomic variables [1]. The study revealed that 77% of the 13/122 reviewed works used two (2) data types, and 23% used three (3) data types [1].

Machine learning and computational intelligence techniques used in technical analysis and hybrid analysis are the hidden Markov model, neural network, neuro-fuzzy inference system, genetic algorithm, time series analysis, regression, mining association rules, support vector machine (SVM), principal component analysis (PCA) and rough set theory, in which the neural network and support vector machine techniques are the most used. The prediction accuracy of models built based on the neural network technique is higher than that of the support vector machine technique [1]. Furthermore, the article also revealed that 92% of the algorithms used were classification machine learning algorithms.

Deep learning is the neural network multilayer model, has been shown to have good pattern recognition. However, deep learning neural network is only suitable for predictive exercises on a dataset where the number of observations is large, but the number of variables is not too large [10, 11]. In contrast, real-world economic-financial datasets often have a small number of observations even much smaller than the number of variables and a large number of variables. The way to overcome this drawback is to use variable dimension reduction techniques for original datasets.

Methodology of predicting the target variable on a large number of original variables, in general, includes two phases [12]. The first phase is to implement some dimensional reduction techniques to transform high dimensional original datasets into lower-dimensional new datasets, but still to preserve important information in the original dataset much as possible. The second phase uses a regression/classification technique on new datasets received in the first phase to build a forecast/classification model of the target variable according to the original variables. The forecast model here implies that it is a predictive model built based on regression techniques.

The most commonly used and most effective dimensionality reduction techniques in economic-financial forecasting exercises on large datasets are the PCA and sparse

PCA methods. Many articles have shown that the dynamic factor model where factors are extracted using the PCA method is superior to other benchmark forecast models in terms of forecast accuracy [13–15] However, the PCA method is only effective for reducing the variable dimension of datasets if their data points are approximately a hyperplane; otherwise, the method is no longer so [16, 17]. Although there are now many different dimensionality reduction methods, it can be said that no dimensionality reduction method has been considered as a natural extension and overcomes the drawback mentioned above of the PCA method.

In our other recent study, we have proposed a kernel tricks-based variable dimension reduction method called KTPCA to distinguish it from the well-known nonlinear kernel PCA dimensionality reduction method [18, 19]. We have shown that: the KTPCA method is another natural extension and overcomes the drawback of the PCA method. Furthermore, the variable dimension reduction performance of the KTPCA method based on an RMSE-best model outperforms the PCA method and the family of SPCA methods, where the variable dimension reduction performance of a method is measured by the RMSE (the root mean squared forecast error) of the model built on the factors extracted from the given dataset by that method.

This article implements a daily forecast of the VN30 stock index on a large dataset of economic-financial predictors by using the dynamic factor model, where factors are extracted from the given input dataset using the KTPCA method based on an RMSE-best model.

This article is structured as follows; following this section, Sect. 2 introduces the used dataset. Section 3 presents the method of building the forecasting model of the VN30 stock index. Some main results and a preliminary analysis of those results are presented in Sect. 4. Section 5 is a few conclusions.

2   Data

Dataset of 60 time-series predictors are collected daily from January 25, 2016, to July 21, 2017. This set includes 390 observation days and is used to implement the forecast of the VN30 index. Table 1 below presents the predictors of that dataset.

Table 1. Variables of the dataset

Data	Variables properties	Sources
Thirty-one variables include the VN30 index and close price of 30 stocks in the basket for VN30 index calculation. Stock code is used as the name of the variable	Reflect domestic factors affecting the VN30 index	HSC Securities Center: www.hsc.com.vn
Twenty-nine variables include the Gold price, the Oil price, and 27 international stock indexes such as S&P 500, NASDAQ, NIKKEI225, Hang Seng (Hong Kong), SENSEX (India), KOSPI (Korea), etc.	Reflect external factors affecting the domestic stock market	Web Finance Yahoo: https://finance.yahoo.com/

Thus, it can be seen that the predictors belong to two data types: past stock data and macroeconomic variables (price predictors). In response to the fact that there are some days of the year in which Vietnam stock market trades whereas some other foreign stock markets do not, the article assumed that each foreign stock index in those days equals the average of this index in the previous nearest day and next nearest day.

The article divides the dataset into two sets, the training dataset includes 385 observation days from January 25, 2016, to July 14, 2017, and the testing dataset includes 05 observation days from July 17, 2017, to July 21, 2017.

3   Method

The VN30 index forecasting method is presented in Fig. 1 below and briefly explained as follows:

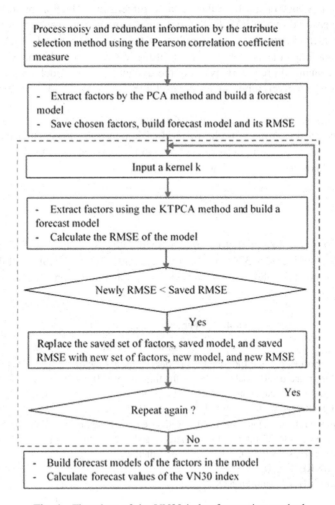

Fig. 1. Flowchart of the VN30 index forecasting method

3.1 Step 1: Eliminate Redundant or Irrelevant Variables to the VN30 Index

Since large datasets of predictors may have noisy or redundant information, and such information must be removed from this dataset first. This article has solved that problem by using the Pearson correlation coefficient [20] to select nonredundant predictors and highly relevant to the VN30 index. Assume Y, X_1, X_2 are time series, Corr (Y, X) is the Pearson correlation coefficient between Y and X; the α, β are user-defined thresholds and are called relevant and redundant thresholds, respectively, X_1 is called relevant with Y if $|\text{Corr } (Y, X_1)| \geq \alpha$; X_2 is called redundant for Y if there exists a variable X_1 so that $|\text{Corr } (Y, X_1)| \geq |\text{Corr } (Y, X_2))|$ and $|\text{Corr } (X_1, X_2)| \geq \beta$. The dataset of the selected predictors is used to replace the dataset of the original predictors in building the VN30 index forecast model. It is called the input dataset and can still be very large. Therefore, it is necessary to perform the variable dimension reduction using the KTPCA method based on an RMSE-best model.

3.2 Step 2: Extract Factors Using the PCA Method, Build a Forecast Model, and Calculate the RMSE of This Model

Assume that the input dataset is centered (i.e., the sum of column vectors is zero). The KTPCA method based on an RMSE-best model is started using the PCA method to extract the principal component factors. Here each principal component factor is a linear projection of the mean-centered input dataset onto an eigenvector of the covariance matrix of the input dataset [21]. P principal component factors corresponding to the P largest eigenvalues of the covariance matrix such that the cumulative eigenvalue percentage on total eigenvalue is greater than a predefined q(%) are chosen to replace the input predictors when building the VN30 index forecast model. Here the dynamic factor model to forecast the VN30 index has the form [22, 23]:

$$Y = \sum_{i=1}^{n} \sum_{j=0}^{n} a_{ij} X_i(-j) + \sum_{k=1}^{q} b_k Y(-k) + c + u \tag{1}$$

where Y (i = 1, 2,..., n) are stationary time series (Y is the VN30 index or a transformation form of this index to ensure that Y is a stationary time series); X_i is a factor extracted by the PCA; u_t is the residual assumed to be white noise, the a_{ij}, b_k, c, are parameters estimated by the Ordinary Least Squares method; $X_i(-j)$ is variable X_i lagged j steps; the r_i (i = 1,..., n), q are maximum lags of factors X_i and Y, respectively. The maximum lag of all variables in a model can be determined by implementing some statistical tests, but if the seasonality length of the variables is known, The maximum lag can be well determined without performing statistical tests. In this article, since the dataset used to build the VN30 index forecast model is collected by the working day of a week (5 days/week), so seasonality length of the input dataset is 5. That implies that the maximum lag of all variables in the model (1) is defined as 5.

3.3   Step 3: The KTPCA Method Based on the RMSE-Best Model

Step 3 is the iterative process of the implementation of the KTPCA method according to kernels. The KTPCA method differs from the PCA method in only that each factor extracted by this method is a linear projection of the mean-centered input dataset onto an eigenvector of the kernel matrix [19]. At the end of each such loop, it is necessary to check if the RMSE of the newly built model is less than the saved RMSE; if so, replace the saved model, the saved RMSE, and the saved set of factors by the newly built model, the RMSE of this model, and the new set of chosen factors, respectively.

It should be noted that although the kernels that can be chosen are very rich, as long as it is a positive definite function, but in practical applications, such kernels are often polynomial kernels $\kappa(x_i, x_j) = \left(<x_i, x_j> + c \right)^d$ or Gaussian kernels $\kappa(x_i, x_j) = \exp\left(-\rho.\|x_i - x_j\|^2 \right)$ or sigmoid kernels $\kappa(x_i, x_j) = \tanh(d_1.\langle x_i, x_j \rangle + d_2)$, where $d \in \mathbb{N}^+$ and $c \geq 0$; $\rho > 0$ [17, 24, 25]. With a polynomial kernels, if d = 1, c = 0, $\kappa(x_i, x_j) = <x_i, x_j>$ then $\kappa(x_i, x_j)$ is the inner product of two vectors x_i, x_j and $K = [\kappa(x_i, x_j)]$ is a covariance matrix of X. In this case, KTPCA and PCA methods are the same. The polynomial and Gaussian kernels are always positive definite functions.

At the end of the iterative process, we get the most suitable kernel among the reported kernels, the VN30 index forecast model having the smallest standard error (or RMSE) among the built models, and the factors in the model. The termination of the variable dimension reduction process using the KTPCA method based on an RMSE-best model is up to the user. The longer the iterative process with suitable kernels, the higher the quality of the built forecast model.

3.4   Step 4: Building Forecast Models of the Factors and Calculate Forecast Value of the VN30 Index

To forecast the VN30 index, we need to forecast the factors using an auxiliary regression model. It can be the ARIMA model, AR (p) model with deterministic trends, or exponential smoothing model [22].

4   Results and Evaluations

4.1   Results

Step 1: Before presenting forecast results, the article introduces some intermediate results. With the relevant and redundant thresholds, respectively, to be 0.05 and 0.85, by using the Pearson correlation coefficient, the article removes 22 irrelevant or redundant variables for forecasting purposes of the VN30 index. Then the dataset of the 59 original variables is replaced by the dataset of 39 original input variables, as shown in Table 2.

Table 2. The relevant and nonredundant original variables of the VN30 index

No	Variable	Pear. corr. coefficient	No	Variable	Pear. corr. coefficient
1	SENSEX	0.9478	21	FPT	0.4431
2	CHINA50	0.8687	22	DPM	0.4167
3	FTSE100	0.8482	23	VCB	0.3931
4	BEL20	0.8468	24	CTG	0.3714
5	RU2000TR	0.8054	25	BMP	0.3436
6	IBEX_35	0.7987	26	CTD	0.3216
7	S_P_ASX_200	0.7903	27	MSN	0.2752
8	NIKKEI225	0.6818	28	SBT	0.2676
9	SSE_COMPOSITE	0.6764	29	KBC	0.2664
10	DHG	0.6700	30	MBB	0.2619
11	HSG	0.6049	31	ITA	0.2249
12	GMD	0.5715	32	GOLD	0.2179
13	HPG	0.5663	33	CII	0.1620
14	REE	0.5616	34	HAG	0.1524
15	BID	0.5573	35	GAS	0.0980
16	KDC	0.5301	36	VNM	0.0764
17	MWG	0.5248	37	VIC	−0.0804
18	OIL	0.5099	38	NT2	−0.1453
19	STB	0.5016	39	PPC	−0.1729
20	SSI	0.4876			

Step 2: With a cumulative eigenvalue percentage threshold of 75%, 12 principal component factors have a cumulative variance percentage of 76.27% extracted from the dataset of the 39 selected original variables using the PCA method. RMSE of the VN30 index forecast model on these 12 factors is shown in the first line in Table 3.

Table 3. The variable dimension reduction results by the KTPCA based on an RMSE-best model

Kernels	Parameters	No of factors	Cum. eigenvalue percentage	RMSE
Polynomial	**d = 1**	**12**	**75.11**	**4.349**
	d = 2, c = 0	2	78.80	4.595
	d = 3, c = 0	1	93.40	4.644
Gaussian	$\rho = \frac{1}{12}e^{-8}$	2	76.50	4.570
	$\rho = e^{-9}$	4	76.30	4.570
	$\rho = \frac{1}{2}e^{-16}$	**10**	**75.91**	**4.328**
	$\rho = \frac{1}{2}e^{-18}$	4	76.37	4.570

Step 3: With each experimented kernel as shown in Table 3, at the end of each loop, the smallest number of chosen factors with cumulative eigenvalue percentage $\geq 75\%$

and RMSE of the forecast model of the VN30 index on the chosen factors are shown in Table 3.

At the end of the variable dimension reduction process, we receive the most suitable kernel to be $\kappa(x_i, x_j) = \exp(\frac{-\|x_i - x_j\|^2}{2.e^{16}})$, 10 chosen factors with the cumulative eigenvalue percentage of 75.91%, the VN30 index forecast model based on an RMSE-best model, and RMSE of this forecast model to be **4.328.** The forecast model of VN30 index is defined as follows:

$$
\begin{aligned}
VN30 = \ &9.70e-08 * PC1(-2) - 7.29e-08 * PC1(-3) + 4.62e-08 * PC1(-4) - 7.80e-08 * PC1\ (-5) \\
&(4.20e-08) \qquad (4.28e-08) \qquad (2.40e-08) \qquad (4.00e-08) \\
&- 2.41e-08 * PC2 - 7.20e-08 * PC2(-2) + 4.69e-08 * PC2(-3) + 5.11e-08 * PC2(-5) \\
&(1.33e-08) \qquad (2.43e-08) \qquad (2.45e-08) \qquad (2.21e-08) \\
&- 5.63e-09 * PC6(-3) - 3.11e-08 * PC9 + 2.98e-08 * PC9(-1) \\
&\qquad\qquad\qquad\qquad\qquad\qquad\qquad\qquad\qquad 3.70e-08 * PC10 + \\
&(2.99e-09)\,(9.10e-09)\,(9.25e-09)\,(9.57e-09) \\
&+ 1.77e-08 * PC10(-1) + 1.46e-08 * PC10(-3) + 1.09 * VN30 \\
&\qquad\qquad\qquad\qquad\qquad\qquad\qquad\quad (-1) - 0.09 * VN30(-2) \\
&(8.96e-09)\,(8.68e-09)\,(0.05)\,(0.005) \\
&R^2 : 0.993; \text{Durbin-Watson stat:} 1.935; \text{SMPL: } 380 \quad \text{after} \quad \text{adjustments}
\end{aligned}
$$

(2)

where PCI (I = 1, 2,..., 10) are factors used to replace the 59 original variables.

Step 4: Since only 05 factors, PC1, PC2, PC6, PC9, and PC10 appeared in the VN30 index forecast model, we only need to build forecast models for these 05 factors. Forecast models of the 05 factors are built based on the model AR(p) with a quadratic trend [26] and are estimated using the OLS regression method. The details of these models are presented in Table 4.

Table 4. Forecast models of chosen factors in the forecast models of the VN30 index

No	Forecast models of factors
1	PC1 = 0.28*PC1(-1) + 0.16*PC1(-3) + 0.09*PC1(-4) + 0.13*PC1(-5) + 3657743.39 - 30.76*t^2
	(0.049) (0.050) (0.053) (0.050) (1043564) (12.43)
	R^2: 0.321; Durbin-Watson stat: 2.004; SMPL: 380 after adjustments.
2	PC2 = 0.37*PC2(-1) + 0.28*PC2(-3) + 10405159.51 - 91.42*t^2
	(0.045) (0.045) (2048002) (23.435)
	R^2: 0.474; Durbin-Watson stat: 2.073; SMPL: 382 after adjustments.
3	PC6 = 0.27*PC6(-1) - 0.14*PC6(-4) + 0.19*PC6(-5) - 151.72*t^2
	(0.083) (0.081) (0.081) (39.83)
	R^2: 0.168; Durbin-Watson stat: 1.932; SMPL: 380 after adjustments.
4	PC9 = 0.37*PC9(-1) - 0.19*PC9(-2) + 0.22*PC9(-3) + 514742761.2 - 3330781.66*t +
	(0.088) (0.093) (0.088) (2.52e08) (1574784)
	+ 5297.47*t^2
	(2428.35)
	R^2: 0.260; Durbin-Watson stat: 1.980; SMPL: 382 after adjustments.
5	PC10 = 0.36*PC10(-1) + 0.19*PC10(-3) + 37673.70*t
	(0.079) (0.078) (12380.53)
	R^2: 0.268; Durbin-Watson stat: 1.934; SMPL: 382 after adjustments

With a forecast horizon of 5, Using the forecast models of the 05 factors in Table 4, we get the forecasts of these factors in the next 05 days from July 17, 2017, to July 21, 2017, in Table 5.

Table 5. Ex-ante forecasts of the 5 factors in the VN30 index

Days	PC1	PC2	PC6	PC9	PC10
7/17/2017	−2303796	−8326838	−3.20E+07	27783170	38196850
7/18/2017	−2372852	−8527150	−3.20E+07	29007370	36723570
7/19/2017	−2442090	−8727987	−3.20E+07	30249040	35023980
7/20/2017	−2511510	−8929348	−3.30E+07	31508180	34393350
7/21/2017	−2581111	−9131234	−3.40E+07	32784790	33923800

Table 6 shows out-of-sample forecast values of the VN30 index. They are calculated based on model (2) and based on the forecast values of the factors. The last column of this table shows the percentage of forecast errors using the built model.

Table 6. Forecast error percentage of the forecast result

Days	VN30	VN30F	Forecast error percentage
7/17/2017	745.59	725.23	2.81
7/18/2017	741.32	725.54	2.17
7/19/2017	737.77	726.53	1.55
7/20/2017	736.64	727.54	1.25
7/21/2017	728.47	728.43	0.005

here, the percentage of forecast errors is defined as the actual value minus the forecasted value divided by the actual value.

4.2 Analysis and Evaluation

Although some important essential tests were not performed to ensure that the estimate of the model (2) is best, linear, and unbiased (BLUE estimate) [26], it can say that the statistical parameters in the forecast model, such as the statistical parameter R^2 and the Durbin-Watson statistic all have high statistical significance [26]. So it can be expected that model (2) has been estimated quite well.

Slightly different from the forecast model of the VN30 index (Model (2)), R^2 in the forecast models of factors presented in Table 4 is relatively small despite the addition of a quadratic trend into these models. Analyzing the statistical characteristics of the 05 principal factors in Table 7 also shows a significant difference between the minimum and maximum values, and the standard deviation of these factors is relatively high. Moreover, the Kurtosis and Jarque-Bera statistics analysis show that except PC2, the probability distribution of the remaining 04 factors is not a normality distribution. That implies that the accurate forecast of factors in model (2) is difficult. So, it is necessary to

detect and process outlier data and use some dummy variables for abnormal observations. This article has not done so. Therefore, it is not guaranteed that the estimate of
the model (2) and the models in Table 4 is BLUE. However, the Durbin-Watson
statistic of these models shows that the models' residuals are not serial correlation [26],
so these models are still used to forecast with acceptable.

Table 7. Statistical characteristics of factors in the model (2)

	PC1	PC2	PC6	PC9	PC10
Mean	6620592	17220017	-1374849	590718.1	1668111
Maximum	36674484	65119864	6.10E+08	1.16E+08	1.45E+08
Minimum	-58929636	-80192860	-91426010	-96035900	-57636750
Std. Dev.	11645746	23203211	81677088	28135076	33474098
Skewness	-0.92	-0.18	2.66	-0.02	1.38
Kurtosis	7.53	3.22	15.03	4.83	5.3
Jarque-Bera	384.08	2.95	2776.55	53.95	208.11
Probability	0	0.228	0	0	0
Observations	385	385	385	385	385

Table 6 shows that the forecast error percentage between the actual VN30 and the
forecasted VN30 (or VN30F) is relatively low, although the model (2) and the forecast
models of factors in Table 4 can be not guaranteed that these models are well estimated.
It can also be seen that while the actual VN30 tends to decrease, the VN30F tends to
increase despite a slight increase. This is very common for point value forecasts even
though the quality of the forecast model is very high because the point forecast value is
only the average of a forecast interval. The main solution to overcome this constraint is
making interval forecasts or combining forecasts using quantitative forecast models
with qualitative analyses, such as using expert analysis to adjust the forecast results
[27], and this article has not implemented such a combination.

5   Conclusions

The VN30 index forecast model is built based on the dynamic factor model, where
factors are extracted from the input dataset of predictors using the KTPCA method
based on an RMSE-best model. The forecast accuracy of this model is quite high,
although, in the building process of the models, some statistical tests have not been
performed and processed if they are violated to ensure that the estimate of the models is
BLUE.

This article shows that the combining the feature selection method using the
Pearson correlation coefficient measure and the KTPCA feature learning method based
on an RMSE-best model increases the effectiveness of reducing the number of predictors and improving the forecast accuracy of the built model.

References

1. Nti, I.K., Adebayo, F.A., Benjamin A.W.: A systematic review of fundamental and technical analysis of stock market predictions. Artif. Intell. Rev. **49**, 3007–3057 (2019)
2. Anbalagan, T., Maheswari, S.U.: Classification and prediction of stock market index based on fuzzy meta graph. Procedia Comput. Sci. **47**(C), 214–221 (2015)
3. Bisoi, R., Dash, P.K.: A hybrid evolutionary dynamic neural network for stock market trend analysis and prediction using unscented Kalman filter. Appl. Soft Comput. J. **19**, 41–56 (2014)
4. Rajashree, D., Dash, P.K., Bisoi, R.: A self-adaptive differential harmony search based optimized extreme learning machine for financial time series prediction. Swarm. Evol. Comput. **19**, 25–42 (2014)
5. Tsai, C.F., Hsiao, Y.C.: Combining multiple feature selection methods for stock prediction: union, intersection, and multi-intersection approaches. Decis. Support. Syst. **50**(1), 258–269 (2010). https://doi.org/10.1016/j.dss.2010.08.028
6. Ghaznavi, A., Aliyari, M., Mohammadi, M.R.: Predicting stock price changes of Tehran artmis company using radial basis function neural networks. Int. Res. J. Appl. Basic. Sci. **10**(8), 972–978 (2016)
7. Agarwal, P., Bajpai, S., Pathak, A., Angira R.: Stock market price trend forecasting using. Int. J. Res. Appl. Sci. Eng. Technol. **5**(IV), 1673–1676 (2017)
8. Talib, R., Hanif, M.K., Ayesha, S., Fatima, F.: Text mining: techniques, applications, and issues. Int. J. Adv. Comput. Sci. Appl. **7**(11), 414–418 (2016)
9. Khan, H.Z., Alin, S.T., Hussain, A.: Price prediction of share market using artificial neural network (ANN). Int. J. Comput. Appl. **22**(2), 42–47 (2011)
10. Kapetanios, G., Papailias, F., et al.: Big data & macroeconomic nowcasting: Methodological review. Econ. Stat. Cent. Excell. Discuss. Pap. Escoe. DP-2018-12, Econ Stat Cent Excell (2018)
11. Hinton, G.E., Salakhutdinov, R.R.: Reducing the dimensionality of data with neural networks. Science (80-). **313**(5786), 504–507 (2006)
12. Chandrashekar, G., Sahin, F.: A survey on feature selection methods. Comput. Electr. Eng. **40**(1), 16–28 (2014). https://doi.org/10.1016/j.compeleceng.2013.11.024
13. Baffigi, A., Golinelli, R., Parigi, G.: Bridge models to forecast the euro area GDP. Int. J. Forecast. **20**(3), 447–460 (2004)
14. Urasawa, S.: Real-time GDP forecasting for Japan: a dynamic factor model approach. J. Jpn. Int. Econ. **34**, 116–134 (2014)
15. Chikamatsu, K., et al.: Nowcasting Japanese GDPs. Bank of Japan Working Paper Series (2018)
16. Van Der, M.L., Postma, E.: Dimensionality reduction: a comparative. J. Mach. Learn. Res. **10**, 13 (2009)
17. Sarveniazi, A.: An actual survey of dimensionality reduction. Am. J. Comput. Math. **04**(02), 55–72 (2014). https://doi.org/10.4236/ajcm.2014.42006
18. Schölkopf, B., Smola, A., Müller, K.-R.: Kernel principal component analysis. In: Gerstner, W., Germond, A., Hasler, M., Nicoud, J.-D. (eds.) ICANN 1997. LNCS, vol. 1327, pp. 583–588. Springer, Heidelberg (1997). https://doi.org/10.1007/BFb0020217
19. Schölkopf, B., Smola, A., Müller, K.-R.: Nonlinear component analysis as a kernel eigenvalue problem. Neural. Comput. **10**(5), 1299–1319 (1998)
20. Guyon, I., Elisseeff, A.: An introduction to variable and feature selection. J. Mach. Learn. Res. **3**(Mar), 1157–1182 (2003)
21. Shlens J.: A tutorial on principal component analysis. arXiv Prepr arXiv14041100 (2014)

22. Koop, G., Quinlivan, R.: Analysis of Economic Data, vol. 2. Wiley, Chichester (2000)
23. Panagiotelis, A., Athanasopoulos, G., Hyndman, R.J., Jiang, B., Vahid, F.: Macroeconomic forecasting for Australia using a large number of predictors. Int. J. Forecast. **35**(2), 616–633 (2019)
24. Kim, K.I., Franz, M.O., Scholkopf, B.: Iterative kernel principal component analysis for image modeling. IEEE Trans. Pattern. Anal. Mach. Intell. **27**(9), 1351–1366 (2005)
25. Schölkopf, B., Smola, A.J.: A Short introduction to learning with kernels. In: Mendelson, S., Smola, A.J. (eds.) Advanced Lectures on Machine Learning, LNCS, vol. 2600, pp. 41–64. Springer, Berlin (2003). https://doi.org/10.1007/3-540-36434-X_2
26. Greene, W.H.: Econometric Analysis, 7th edn. New York University, Prentice Hall, New York (2012)
27. Armstrong, J.S., Green, K.C.: Forecasting methods and principles: evidence-based checklists. J. Glob. Schol. Market. Sci. **28**(2), 103–159 (2018)

An Improved Algorithm to Protect Sensitive High Utility Itemsets in Transaction Database

Nguyen Khac Chien[1(✉)] and Dang Thi Kim Trang[2]

[1] Faculty of Foreign Languages and Informatics, People's Police University, Ho Chi Minh City, Vietnam
[2] Posts and Telecommunications Institute of Technology Ho Chi Minh City, Ho Chi Minh City, Vietnam

Abstract. Privacy-Preserve Utility Mining is becoming a topic of interest to many researchers. The goal is to protect the sensitive-high utility itemsets in the transaction databases from being exploited by data mining techniques. This paper studies methods to hide sensitive high utility itemsets in transaction databases. There are some effective methods to deal with this problem, but these methods still cause undesirable side effects, such as: being missing hidden itemsets with non-sensitive high utility itemsets, the difference between the original database and the modified database, etc. This paper proposed an improved algorithm for hiding sensitive high utility itemsets, called IEHSHUI, focus on choosing the order to hide sensitive itemsets and selecting items to modify with minimal side effects. Experimental results show that the IEHSHUI proposed algorithm is more efficient than existing algorithms in terms of execution time.

Keywords: High utility itemset · Hiding utility itemset · Privacy-preserving utility mining

1 Introduction

Data mining is used to discover knowledge and decision-making information from huge databases [3]. In the business, data is shared between different companies for mutual benefit. However, this carries the risk of exposing sensitive information contained in that databases [10]. Sensitive information can be represented as a set of frequency patterns and high utility patterns with confidentiality. Therefore, the data owner wants to hide this sensitive information before sharing the database with a partner. To solve this problem, privacy-preserving data mining has been proposed and has become an important research direction [1]. Privacy-preserving data mining methods have been applied in various fields, such as in cloud computing, e-health, wireless sensor networks, and location services [9].

The method to hide sensitive information is to modify the original database into a modified one by modifying some items in the original database. Atallah et al. [2] have demonstrated that the problem of hiding sensitive information in the optimal problem is an NP-hard problem and the authors have proposed a data modification algorithm based on a heuristic strategy. There have much research works in this field. However,

P. Cong Vinh and N. Huu Nhan (Eds.): ICTCC 2021, LNICST 408, pp. 95–107, 2021.
https://doi.org/10.1007/978-3-030-92942-8_9

there is currently very little research on the method of hiding sensitive information to protect high utility itemsets in transaction databases based on the utility concept. Besides, the side effects base on non-sensitive information is also of little concern. Therefore, the database modification can protect non-sensitive information and the quality of the original database has been noticed.

Some works have been published to solve this problem such as works [4, 7, 8, 13–15]. Most of these algorithms use data modification methods to efficiently hide sensitive itemsets. However, they still cause unwanted side effects such as hiding non-sensitive information, database corruption before and after the hiding process. This paper will propose a strategy to improve the performance of the EHSHUI algorithm in [4], called the IEHSHUI algorithm. The proposed algorithm focuses on:

(i)   Choose sensitive itemset S_j which has the maximal utility to hide.
(ii) Select item $i \in S_j$, which is among the most sensitive itemsets to modify. If many items that satisfy, then choose the item from the least number of non-sensitive itemsets to modify, called victim item i_{vic}.
(iii) The IEHSHUI proposed algorithm will use the coefficient α in [14] to calculate the ratio of the number of item i_{vic} in all sensitive transactions ST supporting the sensitive itemset $S_j \in ST$ to reduce processing time.

The rest of the paper organized as follows. Section 2 basic concepts and reviews related works. Section 3 proposes an improved hiding high utility item set algorithm. Section 4 experimental results and analysis. Finally, we present the conclusion of the paper in Sect. 5.

2   Basic Concepts and Related Work

2.1   Basic Concepts

In this paper, we use Table 1 and Table 2 as illustrative examples. In Table 1, the database contains nine transactions. In Table 2, the profit table of each item is contained in the transaction database.

Some concepts of high utility itemsets mining are presented in [5].

Let $I = \{i_1, i_2, \ldots, i_m\}$ be a distinct set of m items, where each item $i_p \in I$ has a profit, called $eu(i_p), 1 \leq p \leq m$ and $D = \{T_1, T_2, \ldots, T_n\}$ represents a transaction database, where T_i is a subset of items contained in I. Each transaction is assigned a unique identifier TID. A set containing one or more items is called an itemset. A transaction T supports an itemset X if $X \subseteq I$. An itemset $X = \{i_1, i_2, \ldots, i_k\}$ contain k items is called a k-itemset. Each item i_p in the transaction T_q is associated with a number of item i_p in the transaction T_q, called $iu(i_p, T_q)$.

Table 1 contains a transaction database of a store which have six items a, b, c, d, e, f. There are nine transactions $T_1, T_2, \ldots, T_9$ in this transaction database. Each transaction contains the sold items along with their quantity. For example, T_1 transaction have three items a, b, and e have been sold with corresponding quantities $iu(a) = 10$, $iu(b) = 2$ and $iu(e) = 5$.

Definition 1. Profit of an item i_p represents the importance of an item i_p, denoted by $eu(i_p)$. For example, in Table 2, $eu(b) = 2$ and $eu(d) = 1$.

Definition 2. The utility of an item i_p in a transaction T_q, denoted by $u(i_p,, T_q)$, is calculated as follows:

$$u(i_p, T_q) = iu(i_p, T_q) * eu(i_p) \tag{1}$$

For example, $u(b, T_8) = iu(b, T_8) * eu(b) = 15 * 2 = 30$.

Definition 3. The utility of an itemset X in a transaction T_q, denoted by $u(X,, T_q)$, is calculated as follows:

$$u(X, T_q) = \sum_{i_p \in X} u(i_p, T_q) \tag{2}$$

For example, $u(bd, T_8) = u(b, T_8) + u(d, T_8) = 15 * 2 + 35 * 1 = 65$.

Table 1. Transaction database

TID	Transactions
T1	(a,10),(b,2),(e,5)
T2	(c,4), (d,2), (e,7), (f,15)
T3	(b,15), (c,15), (e,1), (f,1)
T4	(a,5), (b,4), (c,20), (d,2), (e,5)
T5	(b,25), (c,15)
T6	(a,15), (e,7), (f,15)
T7	(a,25), (c,15), (d,40)
T8	(b,15), (d,35), (e,3)
T9	(a,5),(b,10),(c,20),(d,30),(e,2),(f,3)

Table 3. High utility itemsets (HUI) *minutil* = 250

HID	Itemset	Utility
1	ef	425
2	a	422
3	acd	372
4	aef	367
5	ae	342
6	f	340
7	af	322
8	ad	317
9	ac	300
10	cdef	281
11	cef	279
12	def	257

Table 2. Profit table

Item	a	b	c	d	e	f
Profit	7	2	1	1	5	10

Definition 4. The utility of an itemset X, denoted by $u(X)$, is calculated as follows:

$$u(X) = \sum_{X \subseteq T_q \wedge T_q \in D} u(X, T_q) \tag{3}$$

For example, $u(bd) = u(bd, T_4) + u(bd, T_8) + u(bd, T_9) = 10 + 65 + 50 = 125$.

Definition 5. The utility of a transaction T_q, denoted by $tu(T_q)$, is calculated as follows:

$$tu(T_q) = \sum_{i_p \in T_q} u(i_p, T_q) \qquad (4)$$

For example, $tu(T_8) = u(b, T_8) + u(d, T_8) + u(e, T_8) = 30 + 35 + 15 = 80$.

Definition 6. The problem of high utility itemset mining. An itemset X is said to be a high utility itemset if the utility of X is greater than or equal to the minimum utility threshold specified by the user, denoted *minutil*. Let *HUI* the set of high utility itemsets, we have $HUI = \{X | X \subseteq I \land u(X) \geq minutil\}$.

The transaction database is given in Tables 1, Table 2, and the minimum utility threshold *minutil* = 250, the set of high utility itemsets (HUI) are shown in Table 3.

2.2   Research Problem

The research problem is stated as follows:

Given a set of sensitive high utility itemsets (referred to as sensitive itemsets) (*SHUI*) must be hidden, denoted by $SHUI = \{S_1, S_2, \ldots, S_s\}$, where $S_j \in HUI$, $1 \leq j \leq s$. The problem of hiding sensitive itemsets is to modify the original database D to a modified database D' such that the utility of all sensitive itemsets must be less than the minimum utility threshold specified by the user, i.e. $u(S_j) < minutil, vo'ij \in [1, s]$.

Definition 7. The set of sensitive itemsets, where si is a sensitive itemset that needs to be hidden before the database is released, we have $SHUI \subseteq HUI$. Let *NSHUI* a set of non-sensitive high utility itemsets (referred to as non-sensitive itemset), we have $SHUI \cup NSHUI = HUI$.

Definition 8. The set of sensitive transactions, each of which contains at least one sensitive itemset, is denoted by *ST*.

The item that is modified to hide the sensitive itemset is called the victim item (i_{vic}). The transaction containing the victim item is called the victim transaction (T_{vic}).

The data modification process of sensitive itemsets hiding consists of the following three steps:

(i)   Apply high utility mining algorithms on transaction database D to get all high utility itemsets (*HUIs*);
(ii) Define the set of sensitive itemsets *SHUI* based on user requirements;
(iii) Apply algorithm to hide sensitive itemsets to generate modified database D'.

2.3   Related Work

In recent years, privacy-preserving utility mining methods have attracted the attention of many researchers. This problem becomes important because it considers both the quantity and profit of each item in the transaction database to hide sensitive high utility itemsets. For privacy-preserving high utility mining to hide sensitive information (sensitive high utility itemsets) in the database, while ensuring other important information is still provided to the adversary, this problem is considered an optimization problem. Finding transactions and items to modify while optimally hiding sensitive high utility information is a difficult and impossible problem [2]. In 2010, Yeh et al. [15] were the first to propose two heuristic algorithms HHUIF and MSICF to hide sensitive high utility itemsets. The two algorithms select the item with the highest utility to modify for the hiding process. The HHUIF algorithm discards the items with the highest utility. The MSICF algorithm considers the number of conflicts in the hidden process.

Then, some works also proposed algorithms to improve the above two algorithms, such as Vo et al. (2013) [14] proposed an algorithm to improve the HHUIF algorithm in terms of time. Selvaraj et al. (2013) [13] propose an algorithm that improves MHIS in selecting the item in case their utility is the same. The results show that the MHIS algorithm is better than the HHUIF algorithm in terms of HF (hiding failure) and MC (missing cost) side effects. Yun and Kim (2015) [16] propose the FPUTT algorithm to improve the efficiency of the HHUIF algorithm by using a tree structure. The FPUTT algorithm is about 5 to 10 times faster than HHUIF. However, the side effects cause the same as HHUIF. Lin et al. (2015) [6] propose three similar measurements to use as a new standard for assessing side effects in privacy-preserving utility mining. Lin et al. (2016) [7] proposed two algorithms MSU-MAU and MSU-MUI to protect high utility itemsets. Both of these algorithms choose the transactions containing the maximal utility of sensitive itemsets to be hidden. These two algorithms apply the Max-Min property of utility to reduce side effects and increase the speed of data modification compared to HHUIF and MSICF algorithms. Furthermore, the MSU-MIU algorithm is better than the MSU-MAU algorithm due to using the optimal projection in the MSU-MIU. Xuan Liu et al. (2020) [8] propose three heuristic algorithms SMAU, SMIU, and SMSE to hide sensitive itemsets in transaction databases. Transactions that support the smallest number of non-sensitive itemsets are selected as modification transactions. These algorithms use two table structures T-table and HUI-table to reduce the number of database scans. However, these algorithms still cause unwanted side effects during hiding the sensitive high utility itemset.

Trieu et al. (2020) [4] proposed an EHSHUI algorithm to improve the HHUIF algorithm. EHSHUI select victim transaction which supports the sensitive itemset has the maximal utility, and the selection victim item that is in the sensitive itemset to be hidden has little effect on the number of non-sensitive itemsets being hidden by mistake or the item with the least utility. The results show that this algorithm is more efficient than HHUIF and MSICF in terms of side effects and execution time. However, the EHSHUI algorithm in [4] still scans the database many times and takes a long time to select the victim item (Algorithm 2 in [4]).

Most of the above algorithms hide all sensitive itemsets but still cause unwanted side effects. Some algorithms may have to scan the database many times leading to a lot of execution time.

In this paper, an effective strategy of selecting the modified item and modified transaction will be proposed so that all sensitive itemsets can be hidden while minimal side effects on non-sensitive information and can reduce processing time.

3 Proposed Algorithm

3.1 Proposal Basics

The goal of hiding sensitive itemsets to protect privacy is not only to hide all sensitive itemsets but also to minimize side effects on non-sensitive information and the integrity of the original database. The algorithm proposed in this paper is also aimed at this goal.

The method of hiding sensitive itemsets is to modify the original database by deleting or reducing the number of some items so that the utility of the sensitive itemsets falls below the minimum utility threshold.

Most published works focus on identifying: Which transaction is selected to be modified (victim transaction - T_{vic})? and which item is selected (victim item $- i_{vic}$) for modification in the victim transaction (T_{vic})?

In this paper, we focus on: Firstly, when hiding sensitive itemsets, the order in which to choose which sensitive itemsets to hide first affects the hidden process and cause unwanted side effects. In this paper, we propose to choose which sensitive itemset has the maximal utility to be hidden first. Because when hiding these sensitive itemsets, it is possible to hide other sensitive itemsets, then we don't need to hide that sensitive itemset anymore. This is demonstrated in the illustrative example. Therefore, it is possible to increase the efficiency of the hiding process. Secondly, we select the victim item (i_{vic}) that is among the most sensitive itemsets to modify. If there are many items satisfied, we select the item from the least number of non-sensitive itemsets to modify. This minimizes the side effects of non-sensitive information. Thirdly, in most of the published algorithms like [4, 7, 8, 11–13, 15], they modify each transaction one by one. This may increase processing time. In this paper, we use the coefficient α in [14] to calculate the quantity reduction rate of item i_{vic} in all sensitive transactions that support Si need to hide. Then the proposed algorithm will modify all sensitive transactions at the same time. This reduces the number of database scans as well as the time it takes to hide the sensitive itemset.

Let S_j be the sensitive high utility itemset. To hide S_j, the utility of S_j must be reduced by at least:

$$diffu = u\left(S_j\right) - minutil + 1 \tag{5}$$

where $u\left(S_j\right)$ is the utility of sensitive itemset S_j and *minutil* is the minimal utility threshold.

The coefficient α is calculated as follows:

$$\alpha = diu \times \frac{eu\left(i_p\right)}{sum\left(i_p\right)} \tag{6}$$

where $diu = \left\lceil \frac{diffu}{eu(i_p)} \right\rceil$, $sum(i_p)$ is the total utility of item i_p all sensitive transactions supporting S_j.

Definition 9. Identify the victim item (i_{vic}): The item that is among the most sensitive itemsets. If there are many items satisfied, we select the item from the least number of non-sensitive itemsets.

The proposed algorithm is presented in Table 4. The algorithm is implemented as follows: First, S_j sensitive itemsets are sorted in descending order of its utility (line 1). Next, perform an iteration to hide the sensitive itemsets from lines 2–13. Calculate the utility to be reduced for each sensitive itemset (line 3). While $diffu > 0$, find the set of sensitive transactions ST. Identify the item that needs to be modified Ivic according to Definition 9 (line 6). Calculate the rate of reduction of the number of item i_{vic}: α (lines 7–8). Lines 9–12 proceed to modify the number of victim item Ivic in sensitive transactions ST. If α < 1, the quantity of victim item i_{vic} in each ST transaction will be reduced by a ratio of α. If α ≥ 1, it will reduce the number of Ivic to 1, to avoid too much deviation in database structure after modification. Perform until all sensitive itemsets are hidden, then the algorithm ends. The proposed algorithm ensures that all sensitive itemsets are hidden.

3.2 Illustration Example

With the database given in Tables 1 and 2, the minimum utility threshold: *minutil* = 250, we can exploit the set of high utility itemsets HUI presented in Table 3.

Suppose the set of sensitive itemsets to be hidden is $SHUI = \{\langle ae \rangle, \langle ef \rangle, \langle aef \rangle\}$
Line 1: Sort *SHUI* in decreasing order of $u(S_j)$:

$$SHUI = \{ef(425), aef(367), ae(342)\}$$

Line 2: Select $S1 = \langle ef \rangle$ to be hidden
Line 3: Calculate $diffu = u(ef) - minutil + 1 = 425 - 250 + 1 = 176$
Line 4: Find the set of sensitive transactions that support $S1 = \langle ef \rangle$ as:

$$ST = \{T2, T3, T6, T9\}$$

Line 5: diff > 0

Table 4. Improvement EHSHUI algorithm (Called IEHSHUI)

Input:	Original database D; High utility itemset HUI; Sensitive high utility itemsets $SHUI = \{S_1, S_2, \ldots, S_s\}$; Minimal utility threshold $minutil$:
Output: Sanitized database D'.	

1	Sort $SHUI$ in decreasing order of $u(S_i)$;
2	**for each** ($S_j \in SHUI$) **do**
3	$diffu = u(S_j) - minutil + 1$.
4	Find set of sensitive transaction ST support S_i.
5	**while** ($diffu > 0$) **do**
6	Find i_{vic} by *Definition 9*.
7	$diu = \left\lceil \frac{diffu}{eu(i_{vic})} \right\rceil$ //Number of item i_{vic} can be reduced to hide sensitive itemset $S_j \in SHUI$
8	Calculate factor $\alpha = diu \times \frac{eu(i_{vic})}{sum(i_{vic})}$, where $sum(i_{vic}) = \sum_{T_q \in ST} u(i_{vic}, T_q)$
9	**for each** $T_q \in ST$ **do**
10	Modify the quantity of i_{vic}.
11	$iu(i_{vic}) = \begin{cases} iu(i_{vic}) - iu(i_{vic}) \times \alpha & if\ \alpha < 1 \\ 1 & if\ \alpha \geq 1 \end{cases}$
12	Modify $diffu$.
13	Update (D);

Line 6: Find the victim item i_{vic} to modify: There are 2 items e and f. Item e is present in the sensitive itemsets $\langle ae \rangle$, $\langle ef \rangle$, and $\langle aef \rangle$. Item f is present in $\langle ef \rangle$ and $\langle aef \rangle$ sensitive itemsets. So choose item e to modify because it is among the most number of sensitive itemsets.

Line 7: Calculate the items e that must be reduced to hide the sensitive itemset $S1 = \langle ef \rangle$ as : $diu = \left\lceil \frac{diffu}{eu(E)} \right\rceil = \left\lceil \frac{176}{5} \right\rceil = 36$

Line 8: Calculate the coefficient α for item e: $sum(e) = u(e, T2) + u(e, T3) + u(e, T6) + u(e, T9) = 35 + 5 + 35 + 10 = 85$

$$\alpha = diu \times \frac{eu(f)}{sum(f)} = 12 \times \frac{10}{340} = 0.35$$

Line 11: Since $\alpha > 1$, the IEHSHUI algorithm modifies the number of item e in transactions T2,T3,T6,T9 to the value 1.

Line 12: And update the values: The utility of the sensitive itemset $S1 = \langle ef \rangle$, reduced to: $u(\langle ef \rangle) = 425 - 65 = 360$.

$$diffu = 360 - 250 + 1 = 111$$

Line 13: Update the database.

Since $diffu > 0$ keep going back to lines 5, 6. IEHSHUI Algorithm selects item f to modify.

Line 7: Calculate the number of items f that need to be reduced to hide the sensitive itemset $S1 = \langle ef \rangle$ as:

$$diu = \left\lceil \frac{diffu}{eu(F)} \right\rceil = \left\lceil \frac{111}{10} \right\rceil = 12$$

Line 8: Calculate the coefficient α for item f.

$$sum\ (f) = u(f, T2) + u(f, T3) + u(f, T6) + u(f, T9) = 150 + 10 + 150 + 30$$
$$= 340$$

$$\alpha = diu \times \frac{eu(f)}{sum(f)} = 12 \times \frac{10}{340} = 0.35$$

So $\alpha = 0.35 < 1$. Calculate the number of items f that must be reduced in transactions $T2, T3, T6, T9$ as:

The item f that must be reduced in $T2$ is $15 * 0.35 = 5$. The number of item f that must be reduced in $T6$ is $15 * 0.35 = 5$. The item f that must be reduced in $T9$ is $3 * 0.35 = 1$. The number of item f that must be reduced in $T3$ is $12 - 5 - 5 - 1 = 1$. But the number of item f in $T3$ is only 1, so when reducing one item f from transaction $T3$, it is considered to remove item f from transaction $T3$. As a result, $T3$ will no support the sensitive itemset $\langle ef \rangle$. The utility of itemset $\langle ef \rangle$ must be reduced by $u(\langle ef \rangle,\ T3)$ when f is removed from transaction $T3$.

Update the values again: $u(<ef>) = 360 - 5 * 10 - 5 * 10 - 1 * 10 - u(<ef>.T3) = 360 - 110 - 15 = 135 < minutil = 250$. So the item set $S1 = <ef>$ has been hidden successfully.

$diff = -114 < 0$.

Line 13: Update the database.

Do the same to hide sensitive itemsets $S2 = <ae>$ and $S3 = <aef>$. Finally, the IEHSHUI proposed algorithm hides all sensitive itemsets and mistakenly hides five non-sensitive itemsets, which are $\langle f \rangle, \langle af \rangle, \langle cdef \rangle, \langle cef \rangle$ and $\langle def \rangle$.

Thus, in the IEHSHUI proposed algorithm, it is possible to modify many transactions at the same time and quickly hide the sensitive itemset. In part 4, the experiment will compare and evaluate the IEHSHUI proposed algorithm with the EHSHUI algorithm in [4].

4   Experimental Results

The experiments were performed on an Intel ® Core™ i7 computer with 2.00 GHz CPU, 8 GB RAM running on Windows 10. Algorithms are implemented in Java language. The experimental database obtained on the website http://www.philippe-fournier-viger.com/spmf/index.php?link=datasets.php has the following characteristics in Table 5:

Table 5. Experimental datasets

Dataset	Number of transaction	Number of item
Chess	3196	75
Mushroom	8124	120

We add the number of item in each transaction at random the values in the range using uniform distribution and the profit values of each item in the database are also generated randomly.

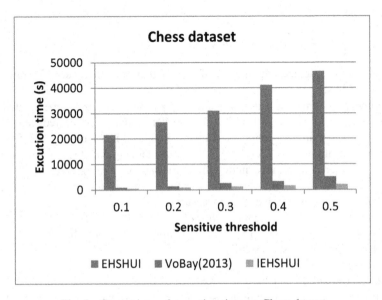

Fig. 1. Comparison of execution time on Chess dataset

In this paper, we compare the IEHSHUI proposed algorithm with the EHSHUI algorithms [4] and the algorithm (VoBay2013) [14] in terms of execution time and memory usage. The experiment was performed fifty times, then we have taken the average value. Experiment on the number of randomly selected sensitive itemsets, 0.1, 0.2, 0.3, 0.4 and 0.5 on the number of high utility itemsets (HUI) respectively. The results are shown in Fig. 1 and Fig. 2.

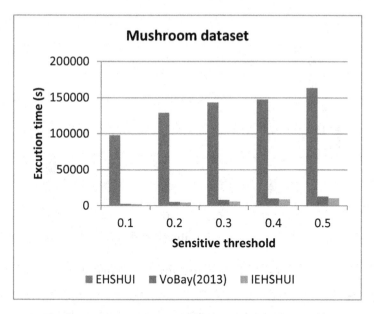

Fig. 2. Comparison of execution time on Mushroom dataset

Figure 1 and Fig. 2 show that the IEHSHUI proposed algorithm is the most efficient in terms of execution time on both Chess and Mushroom dataset. The IEHSHUI
algorithm is faster than the EHSHUI algorithm in [4] many times because the IEH-
SHUI algorithm modifies multiple transactions at the same time to hide sensitive

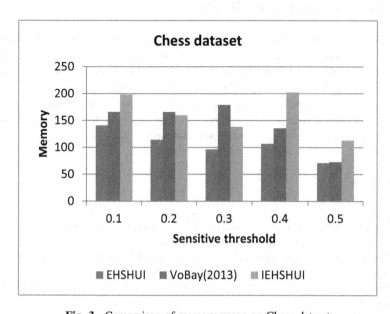

Fig. 3. Comparison of memory usage on Chess dataset

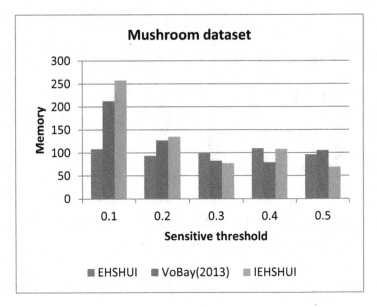

Fig. 4. Comparison of memory usage on Mushroom dataset

information. The EHSHUI algorithm in [4] modifies each transaction one by one. Figure 3 and Fig. 4 show that the memory usage of the IEHSHUI proposed algorithm is more than other algorithms.

5   Conclusion and Future Works

This paper has proposed an algorithm IEHSHUI to protect sensitive itemsets effectively based on the strategy of selecting reasonable sensitive itemsets and victim items. Experiment results show that the IEHSHUI algorithm is more efficient than the EHSHUI [4] and algorithm [14] in terms of execution time.

In the future, we continue to improve the algorithm and test the proposed algorithm on other transaction databases and compare with other algorithms to evaluate the effectiveness and performance on other measurements.

References

1. Agrawal, R., Srikant, R.: Privacy-preserving data mining. In: Proceedings of the 2000 ACM SIGMOD International Conference on Management of Data (2000)
2. Atallah, M., et al.: Disclosure limitation of sensitive rules. In: Proceedings 1999 Workshop on Knowledge and Data Engineering Exchange (KDEX1999) (Cat. No. PR00453). IEEE (1999)
3. Fournier-Viger, P., et al.: A survey of itemset mining. Wiley Interdiscipl. Rev. Data Min. Knowl. Discov. **7**(4), e1207 (2017)

4. Huynh Trieu, V., Le Quoc, H., Truong Ngoc, C.: An efficient algorithm for hiding sensitive-high utility itemsets. Intell. Data Anal. **24**(4), 831–845 (2020)
5. Krishnamoorthy, S.: Pruning strategies for mining high utility itemsets. Expert Syst. Appl. **42**(5), 2371–2381 (2015)
6. Lin, C.-W., et al.: A GA-based approach to hide sensitive high utility itemsets. Sci. World J. **2014** (2014)
7. Lin, J.C.-W., et al.: Fast algorithms for hiding sensitive high-utility itemsets in privacy-preserving utility mining. Eng. Appl. Artif. Intell. **55**, 269–284 (2016)
8. Liu, X., Wen, S., Zuo, W.: Effective sanitization approaches to protect sensitive knowledge in high-utility itemset mining. Appl. Intell. **50**(1), 169–191 (2019). https://doi.org/10.1007/s10489-019-01524-2
9. Mendes, R., Vilela, J.P.: Privacy-preserving data mining: methods, metrics, and applications. IEEE Access **5**, 10562–10582 (2017)
10. O'Leary, D.E.: Knowledge discovery as a threat to database security. Knowl. Discov. Database **9**, 507–516 (1991)
11. Rajalaxmi, R., Natarajan, A.: Effective sanitization approaches to hide sensitive utility and frequent itemsets. Intell. Data Anal. **16**(6), 933–951 (2012)
12. Saravanabhavan, C., Parvathi, R.: Privacy preserving sensitive utility pattern mining. J. Theor. Appl. Inf. Technol. **49**(2) (2013)
13. Selvaraj, R., Kuthadi, V.M.: A modified hiding high utility item first algorithm (HHUIF) with item selector (MHIS) for hiding sensitive itemsets. Int. J. Innov. Comput. Inf. Contrl. **9**, 4851–4862 (2013)
14. Vo, B., et al.: An efficient method for hiding high utility itemsets. In: KES-AMSTA (2013)
15. Yeh, J.-S., Hsu, P.-C.: HHUIF and MSICF: Novel algorithms for privacy preserving utility mining. Expert Syst. Appl. **37**(7), 4779–4786 (2010)
16. Yun, U., Kim, J.: A fast perturbation algorithm using tree structure for privacy preserving utility mining. Expert Syst. Appl. **42**(3), 1149–1165 (2015)

Nowcasting Vietnam's RGDP Using a Kernel-Based Dimensional Reduction Method

Thanh Do Van[✉]

Faculty of Information Technology, Nguyen Tat Thanh University,
Ho Chi Minh City, Viet Nam
dvthanh@ntt.edu.vn

Abstract. The gross domestic product growth rate (RGDP for short) is one of the most important macroeconomic indicators often used for making economic policies and planning production and business development plans by government agencies and enterprise communities. Research to improve the forecast accuracy of this indicator has always been of interest to researchers. In Vietnam, this indicator is only released quarterly.

It is no longer appropriate to forecast the RGDP according to predictors at the same frequency as this indicator because, in the time interval between two quarters, there may be some political, socio-economic events occurring that have a substantial impact on many economic activities that cause the change of the RGDP in the current quarter and the next quarters. It is necessary to use another new forecast approach to overcome this limitation.

The purpose of this article is to build a model to nowcast the RGDP on a large dataset of predictors at higher frequencies than the quarterly frequency. Such a model is developed based on the dynamic factor model. Unlike previous studies, the factors in the built model are extracted from the input dataset by a variable dimension reduction method using kernel tricks and based on an RMSE-best model. The article also proposes a ragged-edge data handling method and reinforcement learning method, suitable for the regression method used to build the nowcasting model of the RGDP indicator.

Keywords: Data mining · Nowcasting · Big data · Dimensionality reduction · Kernel tricks · PCA

1 Introduction

The estimation of Vietnam's RGDP is only performed at quarterly and yearly frequencies. In Vietnam, the RGDP indicator, especially at the quarterly frequency, is now considered one of the most important macroeconomic indicators in the formulation of the economic regulation policies of the government. In the context of deep, broad international economic integration and unpredictable international changes, the forecast of RGDP and the update of forecasts of RGDP according to real-time data flows are essential and have very high practical significance. The nowcasting model of the RGDP indicator makes that possible.

© ICST Institute for Computer Sciences, Social Informatics and Telecommunications Engineering 2021
Published by Springer Nature Switzerland AG 2021. All Rights Reserved
P. Cong Vinh and N. Huu Nhan (Eds.): ICTCC 2021, LNICST 408, pp. 108–128, 2021.
https://doi.org/10.1007/978-3-030-92942-8_10

Nowcast is defined as the prediction of the present, the very near future, and the very recent past by using available, timely, and reliable information to formulate predictions for target variables of interest [1–4]. Nowcast aims to exploit the information published early and possibly at higher frequencies than the target variable of interest to obtain an early estimate before the official figures become available. A nowcasting model of macroeconomic indicators needs to have features to monitor many data releases, forming expectations about them and revising the assessment on the state of these indicators whenever realizations diverge sizeably from those expectations. The nowcasting approach is related to the datasets known as big data [2–5] and ragged-edge data [7–9].

Big Data has been variously defined in the literature. The work [2] introduced three ways to identify big data. First, big data is identified through several characteristics, of which the most important features are the four "V" [2, 9]: Volume (the quantity of generated and stored data), Variety (the type and nature of the data), Velocity (the speed at which the data is generated and processed to meet the demands and challenges), Veracity (the data quality and the data value). Big data can be identified by the number of variables and the number of observations. Then a dataset is called large if either the number of variables or the number of observations or both are large [2, 5]. Big data can also be identified through the content of the data. It is data from social networks, traditional business systems, or the internet of things [2].

There have been many studies on building the RGDP indicator nowcasting models or nowcasting this indicator. These models are built based on the factor state-space model (another version of the dynamic factor model developed according to the idea of Kalman filter) [10, 11], factor bridge equation model [12–14], and factor MIDAS model [12–15]. The factors are extracted from the predictors' dataset using the Principal Component Analysis (PCA) method in these models. The authors of the works [13, 16–18, 24] reviewed methodologies and econometric techniques to forecast macroeconomic indicators on large data sets. They are classified into two categories: statistical machine learning techniques and artificial intelligence machine learning techniques. Techniques based on regression such as the bridge equation, MIDAS, and logistic regression belong to the statistical technical group. The machine learning techniques such as artificial neural network (ANN), support vector machine (SVM), genetic algorithm, cluster analysis, k closest neighbors,... belong to the artificial intelligence technical group, in which the ANN and SVM techniques are the most common algorithms utilized for the forecast/classification purpose [19–22]. One found that ANN's ability in prediction is significantly better than SVM [20]. An essential aspect of deep models is that they can extract rich features from the raw data and make predictions. So, from this point of view, deep models usually combine both phases of feature extraction and prediction in a single phase. ANNs with different structures were tested, and the experiments proved the superiority of deep ANNs over shallow ones.

According to [2], the deep learning neural network method [23] is only appropriate in prediction exercises on datasets with a large number of observations and not a large number of variables. In other words, this method is not suitable for datasets with a large number of variables.

The essence of nowcasting economic-financial indicators is to forecast a target variable (or dependent variable) at a low frequency on a large dataset of time series predictors (or explanatory variables) at some higher frequencies. The works [2, 4, 5,

16–18, 24, 25] showed that the effective modeling method on mixed frequency macroeconomic big data is to use the dynamic factor model and the Kalman filter, in which the dynamic factor model is used more. The dynamic factor model includes the factor bridge equation model and the factor MIDAS model [13, 14, 17, 18], here the factors are extracted from the dataset of input predictors by the PCA method.

The bridge equation model approach [26] offers a convenient solution for filtering and aggregating variables characterized by different frequencies. However, aggregation may lead to the loss of valuable information [17]. This issue has led to the development of a mixed frequency modeling approach called MIDAS [27].

Bai, Ghysels, and Wright [28] studied the relationship between the MIDAS regression and Kalman filter when forecasting mixed frequency data. The authors examine how the MIDAS regression and the Kalman filter match under ideal cases, where the stochastic components, the lag of high and low-frequency variables are all assigned values exactly. The experimental results show that the forecast accuracy of the models built based on the Kalman filter and the MIDAS model is similar. In most cases, the Kalman filter gives a slightly more accurate result, but it requires much more computation. Ankargren, Sebastian, and Unn Lindholm [29] experimented and concluded that the MIDAS model and the bridge equation model achieve a lower forecast error than that of the factor state-space model (it is another version of the dynamic factor model where one used some equations containing factors at a high frequency in a regression model of variables at a low frequency) [11]. That article also showed that the bridge equation model using a small set of variables (≤ 6 variables) performs better than using a medium set of variables (around 14 variables) or large (about 34 variables). The best performance belongs to the MIDAS model when using a set of variables of medium size. However, this article has not shown that with a small set of predictors and in ideal cases, of the two models, the bridge equation model and the MIDAS model, which one has the small RMSE than.

The task of dimensionality reduction techniques/methods is to extract factors from an input data set to replace the predictors in this set. In economics and finance, the most commonly used factor extraction technique is the PCA and sparse PCA. The PCA method is a typical unsupervised learning method to transform a dataset in a high dimensional space into a much lower dimensional space while still preserving the maximum variance and covariance structure of the original dataset [30]. The dataset in the low dimensional space is the dataset of some principal component factors. Each factor is a linear projection of the mean-centered input dataset into an eigenvector of the covariance matrix created from the input dataset of predictors. The cumulative variance percentage of p first factors corresponding to p highest eigenvalues is also the percentage of information of the original dataset held by these p factors. In practice, one usually takes only p factors so that its corresponding cumulative variance percentage is in the range of 70%–90% to replace the original preditor variables.

The PCA method is very efficient for reducing the dimensionality of a dataset [31] if its data points are approximately a hyperplane and not efficient if that is not true. In our recent study, we have proposed a new variable dimension reduction method as a natural extension of the PCA method and called the KTPCA method. With nowcasting models built based on the dynamic factor model, the experiment of the variable dimension reduction methods PCA, SPCA, randomized SPCA, Robust SPCA, and the KTPCA methods on 11 large datasets of time series predictors showed that the performance of the KTPCA method based on an RMSE-best model is always higher than that of the reported methods. Here the performance of a variable dimension reduction method is measured by the RMSE of a nowcasting model built based on the dynamic factor model, in which the factors are extracted using this variable dimension reduction method. Moreover, the dynamic factor model includes the factor bridge equation model, factor unrestricted MIDAS model (U-MIDAS), and factor restricted MIDAS models with the MIDAS weights to be Step, polynomial Almon, and exponential Almon functions. This study also compared the forecast accuracy of the factor bridge equation model, the factor U-MIDAS model, and some other factor-restricted MIDAS models. We received that the forecast accuracy of the factor U-MIDAS model is higher than that of other factor MIDAS models. With the number of factors being small (<=6), the factor bridge equation and U-MIDAS models are competitive in forecast accuracy.

This article applies the results of our recent research just mentioned to build a nowcasting model of Vietnam quarterly RGDP so that the forecast accuracy of the built model is the highest compared to models built based on other known so far. Based on the specific dataset to build the nowcasting model of the RGDP indicator, the nowcasting model is built based on the factor bridge equation model or the factor U-MIDAS model. Here, factors are extracted using the KTPCA method based on an RMSE-best model.

The article is organized as follows: following this section, Sect. 2 introduces some necessary content used in the following sections. Section 3 presents a dataset used to build and a method of building the Vietnam RGDP nowcasting model. Section 4 introduces some main results. Section 5 presents a ragged-edge data handling method and a reinforcement learning method and uses the built model to update the RGDP forecasts in a real-time data flow. The last Sect. 6, is some conclusions.

2 Preliminaries

2.1 Factor Bridge Equation Model

Bridge equations are linear regression ones that link variables at high frequency to lower frequency variables. This method allows early estimates of low-frequency variables by using higher frequency variables [2, 14, 17, 18].

Factor bridge equation model is defined as follows:

$$y_t^Q = \alpha + \sum_{i=1}^{N} \beta_i x_{i,t}^Q + \sum_{i=1}^{K} \gamma_i F_{i,t}^Q + \varepsilon_t \tag{1}$$

where y_t^Q is the RGDP indicator at the quarterly frequency Q; t is the time point of frequency Q; $x_{i,t}^Q$ are the indicators at the same frequency as the target variable y_t^Q; $F_{i,t}^Q$ are factors at the frequency Q aggregated from factors at higher frequencies $F_{i,t}^M$ and/or

$F_{i,t}^D$ (M, D are monthly or daily frequency, respectively), in which $F_{i,t}^M$ or $F_{i,t}^D$, respectively, are extracted from large sets of predictors $z_{i_j,t}^M$ and/or $z_{i_h,t}^D$ by using the KTPCA based on an RMSE-best model.

Due to the economic system has inertia, macroeconomic variables usually exist autocorrelation, so it is necessary to add a corresponding lag in a forecasting model of macroeconomic indicators, so the factor bridge equation model can be further extended to include lags of the target variable as well as of the predictors. Then the Eq. (1) can be written in the form:

$$y_t^Q = \sum_{k=1}^{q} b_k y_{t-k}^Q + \sum_{i=1}^{N} \sum_{j=0}^{r_i} \beta_{ij} x_{i,t-j}^Q + \sum_{j=1}^{P} \sum_{q=0}^{P_j} \gamma_{jq} F_{i,t-h}^Q + c + u_t \quad (2)$$

where r_i (i = 1, ... , n), P_j (j = 1, ... , m) and q, are the maximum lag of the variables $x_{i,t}^Q$, $F_{i,t}^Q$, and y_t^Q, respectively. The maximum lag can be determined using either AIC or BIC information criterion.

Model (2) can be rewritten as:

$$\psi(L)y_t^Q = \sum_{i=1}^{N} \beta_i(L)x_{i,t}^Q + \sum_{j=1}^{M} \gamma_j(L)F_{i,t}^Q + c + u_t \quad (3)$$

where L denotes usual lag operator, $\psi(L) = 1 - \sum_{k=1}^{q} b_k L^k$, $\beta_i(L) = \sum_{j=0}^{r_i} \beta_{ij} L^j$, and $\gamma_j(L) = \sum_{h=0}^{P_j} \gamma_{j_h} L^h$.

In essence, model (2) is also the Autoregressive Distributed Lag ARDL(q, $r_1, ..., r_N, p_1, ..., p_M$) model [32].

2.2   The Factor MIDAS Models

The factor MIDAS model under consideration is [33]:

$$\psi(L)y_t^Q = \sum_{i=1}^{N} \beta_i(L)x_{i,t}^Q + f(\{F_{t/S}^M\}, \theta, \lambda) + u_t \quad (4)$$

$\{F_{t/S}^M\}$ is set of the factors extracted from a large set of predictors sampled at a higher frequency with S values for each low-frequency value; $\psi(L) = 1 - \sum_{k=1}^{q} b_k L^k$; $\beta_i(L) = \sum_{j=0}^{r_i} \beta_{ij} L^j$; f is a function describing the effect of the higher frequency data in the low-frequency regression; $b = (b_k)$, $\beta_i = \left(\beta_{ij}\right)$, θ, and λ are vectors of parameters to be estimated.

If we like only to include each of the higher frequency components as a predictor in the low-frequency regression, then model (4) can be given by

$$\psi(L)y_t^Q = \sum_{i=1}^{N} \beta_i(L)x_{i,t}^Q + \sum_{\tau=0}^{k-1} F_{(t-\tau)/S}^{M}{}^T \theta_\tau + u_t \quad (5)$$

where $F^M_{(t-\tau)/S}{}^T$ are the factors at the τ high-frequency periods before t. Then, a distinct θ_τ is associated with each of the k high-frequency lag factors. The number of θ_τ coefficients may be very large. If these coefficients are not constrained, then model (5) is called the unrestricted MIDAS model (or U-MIDAS). So, the U-MIDAS model offers the greatest flexibility but requires large numbers of coefficients.

2.3   The KTPCA Based on an RMSE-Best Model

Assume that the missing and outlier data in the dataset of the y^Q_t, $x^Q_{i,t}$, and $z^M_{j,t}$ variables have been processed. Without loss of generality, it can assume that all variables y^Q_t, $x^Q_{i,t}$, and $z^H_{j,t/S}$ are stationary time series. Improved from the KTPCA variable dimension reduction method based on an RMSE-best model for data sampled at the same frequency, the KTPCA method based on an RMSE-best model on mixed frequency datasets is shown in Fig. 1 below.

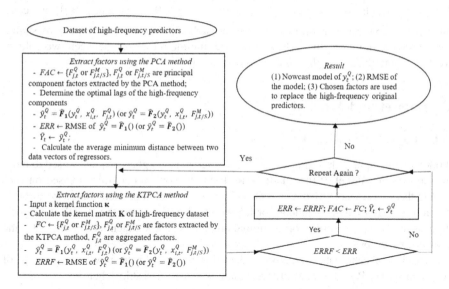

Fig. 1. The KTPCA method based on an RMSE-best model on mixed frequency datasets

In this flowchart, we denote $\hat{F}_1()$ and $\hat{F}_2()$ as nowcasting models of the dependent variable y^Q_t to be estimated based on the factor-bridge equation model (3) and the factor U-MIDAS model (5), respectively; FAC is the set of factors extracted from the dataset of high-frequency predictors using the KTPCA method based on an RMSE-best model where $F^M_{j,t/S}$ are factors extracted from the dataset of high-frequency predictors and are sampled at a higher frequency with S values for each low-frequency value; moreover; $F_{j,t}^Q$ are aggregated from $F_{j,t/S}^M$ at the same low-frequency as the dependent variable. ERR is the standard mean forecast error of a nowcasting model and is measured by the root mean squared forecast error (RMSE for short) determined as follows:

$$\text{RMSE} = \sqrt{\frac{1}{T} \cdot \sum_{t=1}^{T} \left(y_t^Q - \hat{y}_t^Q \right)^2} \tag{6}$$

here, $\hat{y}_t^Q$ is produced by the built nowcasting model and is called a fitted variable of y_t^Q; T is the number of observations. The process above is iterative, and in principle, we can choose a kernel so that the RMSE of its corresponding nowcasting model is small as possible. The above flowchart shows that the most suitable factor extraction and building a nowcasting model of a target variable are combined in one process called the factor extraction using the KTPCA method based on an RMSE-best model.

In practical applications, polynomial kernels $\kappa(x_i, x_j) = \left(c_1 \langle x_i, x_j \rangle + c_2 \right)^d$ and Gaussian kernels $\kappa(x_i, x_j) = \exp(-\frac{\|x_i - x_j\|^2}{2 \cdot \rho^2})$, here $c_1 > 0, c_2 \geq 0$ and $\rho > 0$, are commonly used [34, 35]. According to some researchers, for Gaussian kernels, ρ^2 should be chosen around the value to be the average minimum distance between two column vectors of the dataset of the input predictors [36].

3   Dataset and Method

3.1   Dataset of Original Predictors

Following the approach of this article, the predictors at a higher frequency used to build the nowcasting model of the quarterly RGDP indicator are determined based on economic theories and as broad as possible. So, collected data of predictors used to build the model may contain redundant or irrelevant information for the RGDP indicator's forecasting. In other words, some original predictors can be potential predictors for the nowcasting purpose.

The list of candidate predictors collected data to build the nowcasting model of Vietnam's RGDP at the quarterly frequency is shown in Table 1 below.

Table 1. List of candidate predictors and selected predictors

Indicators	Units	Freq.	The number	Release dates	Source	The number of selected variables
The Gross Domestic Product Growth Rate (RGDP)	%, YoY	Q	01	A[1]	GSO[2]	Target variable
Total retail sales of consumer goods and services	Bil. VND, at current prices	M	01	A	GSO	01
Retail sales in some economic sectors	-	M	04	A	FiinPro[3]	04
Basic inflation, overall CPI, and CPI of 8 other main consumer goods baskets	%, YoY	M	10	A	GSO	08
Gold & US dollar price indices	%, YoY	M	02	A	GSO	0
Consumption index for the whole industry and 18 major production industries	%, YoY	M	19	A	FiinPro	16
The world price of Vietnam rice & Thailand rice; the world price of copper, coffee & rubber	USD/ton at current prices	M	5	B[4]	Fred[5]	04
The world price of Brent crude	USD/Cart on	M	1	B	Fred	01
Total imports & exports of goods and services	Mil. USD, at current prices	M	2	A	GSO	0

(*continued*)

Table 1. (*continued*)

Imports of 18 production industries	-	M	18	A	FiinPro	10
Exports of 25 production industries	Mil. USD, at current prices	M	25	A	FiinPro	14
Inventory index of the whole industry and 18 major production industries	%, YoY	M	19	A	FiinPro	08
Money M2, Deposits of financial institutions and residents	Bil. VND, at current prices	M	03	A	FiinPro	03
Total outstanding loans of the whole economy and five economic sectors of level 1	%, YoY	M	06	A	FiinPro	01
FDI includes implemented, registered, newly registered & additionally registered FDI	Mil. USD, at current prices	M	04	A	FiinPro	04
Industrial production index of the whole economy & of 9 main industries	%, YoY	M	10	A	GSO	08
Deposits from business organizations and residents	%, YoY	M	01	A	FiinPro	0
Manufacturing Purchasing Managers Index (PMI)	Point	M	01	C[6]	IHS-Markit[7]	01
Investment capital implemented from the State budget	Bil. VND, at current prices	M	01	A	GSO	0
Short-term and medium-term lending rates of state-owned commercial banks	%,	W	02	D[8]	FiinPro	0
Interest rates for short, medium, and long-term mobilization of state-owned commercial banks	%,	W	03	D	FiinPro	03

(*continued*)

Table 1. (*continued*)

Exchange rate of VND with USD	Nominal exchange rate	Day	01		FiinPro	0
Exchange rate of Yuan with USD	Nominal exchange rate	Day	01		Fred	01
Vietnam stock indexes: VN Index & HNX index	Point	Day	02		Cophieu 68[9]	01
Down Jones and S&P 5000 stock indexes	Point	Day	02		Fred	01

[a]A: within the last five days of a quarter
[b]https://www.gso.gov.vn/
[c]http://fiinpro.com/
[d]B: within the last three days of a month
[e]http://fred.stlouisfed.org/
[f]C: within the first four days of a subsequent month
[g]https://www.markiteconomics.com/
[h]D: the last day of the working week
[i]http://www.cophieu68.vn

Thus, there are 143 original predictors used to build the nowcasting model of Vietnam's RGDP at the quarterly frequency. The candidate predictors reflect the demand-side, supply-side, and market liquidity of the economy.

Data of variables at the monthly and quarterly frequencies are mainly collected from two sources (Table 1): General Statistics Office (GSO) and FiinPro - the company providing economic and financial data services. These data are usually released during the last five days of every month. The thing to note is data on total export and import, total retail sales of consumer goods and services, consumption index, and inventory index of the whole economy at a monthly frequency are released by GSO, but the FiinPro also releases the mentioned above and their more detailed data. With the same economic-financial indicators, FiinPro's data release date is usually about 2–3 days behind the General Statistics Office.

Interest rates on deposits and loans are released weekly on the first day of working weeks. Survey data on PMI conducted by IHS Markit is usually released on days 1 to 4 of the following month. Data collected from the Fed (stock indices, exchange rates,...) and Cophieu68 (stock indices) are daily figures released before the market closes. Besides, starting from 1/2013, the national account of Vietnam's economy is calculated according to the new economic subsectors and the base year of 2010, while the statistical data of previous years have not been adjusted accordingly. That means that it should only collect data from Q1, 2014 to Q1, 2020 for the RGDP indicator and other predictors at the same frequency as the RGDP, while data of other predictors at higher frequencies can be collected earlier. Specifically, in this article, data at monthly frequency was collected from January 2013 to March 2020. Data of the remaining variables were collected accordingly at the daily or weekly frequencies from January 1, 2013, to March 31, 2020.

3.2   Method

Figure 2 below briefly describes the process of building a nowcasting model of the RGDP indicator based on the most optimal model in the factor bridge equation model and the factor U-MIDAS model. Here, the factors are extracted using the KTPCA method based on an RMSE-best model.

The main content of the original dataset pre-processing is to add missing data, deal with outlier data and deal with the seasonality of the data. The missing or outlier data is overcome by the AR interpolation or smoothing method depending on the characteristics of each specific time series variable. It can be seen that the values of the dependent variable and other predictors in Table 1 include absolute and relative numerical values, in which relative numerical values (%) are compared with the same period last year. Macroeconomic data has often seasonality. The use of relative numerical values for economic variables implies dealing with the seasonality of its data. Predictors receiving absolute numerical values in Table 1 are converted into predictors of the same name that receive relative values compared to the same period last year. Assuming X_t is an absolute numeric value economic variable in Table 1, XS_t is the seasonally processed variable of X_t, the formula that converts the absolute numeric values of X_t into its relative numeric values is determined by:

$$XS_t = LOG(X_t/X_{t-12}) \quad \text{or} \quad XS_t = LOG(X_t/X_{t-4}) \tag{7}$$

That depends on the indicator X_t at the monthly or quarterly frequency, respectively. When X_t is the GDP indicator, then the XS_t is called the RGDP indicator. To convert the absolute numerical values of financial predictors such as price, stock indices, and interest rates at daily or weekly frequency, we must aggregate these predictors at the monthly frequency using the arithmetic average formula and then apply the formula (7). The data conversion in the above way often makes non-stationary economic variables into stationary series (although this is not always the case), factors extracted by the KTPCA method based on an RMSE-best model to be stationary series, and avoids spurious regression when building nowcasting models using that variable dimension reduction method. This data conversion also facilitates the develop of software nowcasting automatically economic and financial indicators. Thus, the RGDP indicator has the observations from Q1/2014 to Q1/2020, while predictors at other frequencies have the observations from January 2014 to March 2020.

The converted dataset may still contain redundant and noisy information for the RGDP indicator's nowcasting purpose, so such information should be eliminated. The method of removing noisy or redundant information is to remove irrelevant or redundant predictors for nowcasting purposes using the Pearson correlation coefficient measure. Here the concepts of relevance and redundancy are defined as follows: suppose Y is the target variable at a low frequency and X is a higher frequency predictor. The predictor X is called redundant or irrelevant to Y if the aggregated variable of X at the same frequency as Y is redundant or irrelevant to Y, respectively. The removal of redundant or irrelevant predictors in such a way is essentially a way of selecting valuable predictors based on the filter approach using the Pearson correlation coefficient measure. The dataset of the selected predictors is called the input dataset.

Calculating the average minimum distance between two column vectors in the input dataset facilitates selecting a suitable Gaussian kernel when implementing the KTPCA method based on an RMSE-best model for extracting factors.

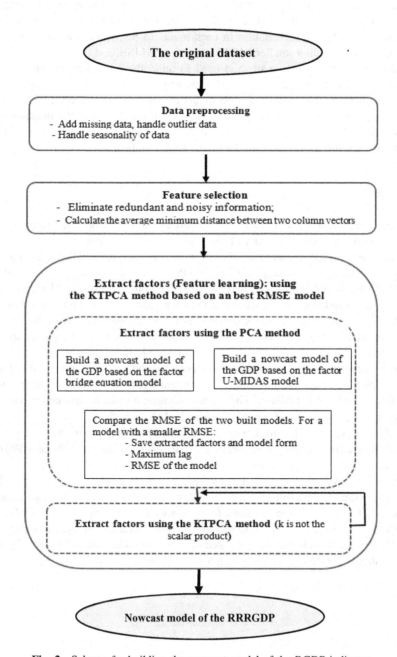

Fig. 2. Schema for building the nowcast model of the RGDP indicator

The extraction of factors from the input dataset of predictors at a frequency higher than the frequency of the target variable using the KTPCA method based on an RMSE-best model is started using the PCA method. On the chosen factors, two RGDP indicator nowcasting models are built based on the factor bridge equation model and the factor U-MIDAS model. Both models are estimated under an ideal condition, which is the maximum lag of all variables in each model to be precisely determined. For a nowcasting model with a smaller RMSE, save it, its RMSE, and the chosen factors. At the same time, the regression method used to build the nowcasting model with the smallest RMSE is also selected to build nowcasting models in the next stage.

The next stage is an iterative process of factor extraction using the KTPCA method, where kernels **k** are not the inner product of two column vectors in the mean-centered input dataset. The details of this process are shown in Fig. 1. At the end of the iterative process, we get the nowcasting model of the RGDP indicator.

4 Results

In this section, the article introduces some intermediate results obtained in the schema for building the nowcasting model of the RGDP indicator in Fig. 2.

With relevant and redundant thresholds of, respectively, 0.1 and 0.9, we get a set of 89 non-redundant and relevant predictors with the RGDP nowcasting purpose as shown in the last column of Table 1, in which the Manufacturing Purchasing Managers Index (PMI for short) and Vietnam Stock index (VN for short) are very highly relevant with the RGDP indicator. The article separates these indices from the set of the monthly frequency predictors and treats them as predictors at the quarterly frequency to facilitate monitoring and assessment of their impact on the RGDP. Their absolute numerical values at each quarter are the arithmetic average of the absolute numerical values at days or months in that quarter. Dataset of 87 remaining predictors at monthly frequency is considered as the input dataset. The average minimum distance between two column vectors of the mean-centered input dataset is $4.32226 = e^{1.463778}$.

Performing the ADF test for the RGDP, VN, and PMI at quarterly frequency shows that the RGDP is the first-order differential stationary while the VN and PMI are stationary series.

With the cumulative eigenvalue percentage threshold of 75%, using the PCA method to extract factors from the centered input dataset, we get the first two principal component factors with the commutative eigenvalue percentage of 77.496%. The results of building the nowcasting model of the RGDP indicator on the two indices VN and PMI and the two chosen factors based on the regression models (3) and (5) above are indicated, respectively, in Tables 2a and 2b below.

Table 2. The RGDP nowcasting model built based on the two dynamic factor models

2a. Nowcasting model is built based on the factor bridge equation model

D(RGDP) = -1.184*D(RGDP(-1))*** - 1.444*D(RGDP(-2))*** - 1.254*D(RGDP(-3))*** -
　　　　　(0.101)　　　　　　　(0.122)　　　　　　　　(0.126)
0.776*PMI*** - 0.324*PMI(-1)*** + 0.076*PMI(-3)* + 0.162*VN*** + 0.089*VN(-1)***
(0.074)　　　　　(0.0438)　　　　　　　(0.0358)　　　　　(0.018)　　　　(0.013)
+ 0.080*VN(-3)***- 0.048*PC1*** - 0.026*PC1(-1)*** - 0.031*PC1(-2)*** + 0.012*PC2***
　(0.011)　　　　　(0.004)　　　　　(0.006)　　　　　　(0.006)　　　　　　(0.002)
- 0.027PC2(-3)*** + 0.005*PC2(-2)*** + 0.017***.
　(0.003)　　　　　(0.001)　　　　　(0.002)
R^2: 99.11; DW stat: 2.40; SMPL: 20 after adjustments; PCI (I =1, 2) are the chosen factors.
　　Significant codes: 0.0001: '****'; 0.001 '***'; 0.01 '**'; 0.05 '*'; 0.1: '.'
- RMSE (dynamic forecast): 0.000796;
- The common maximum lag for all variables: 03.

2b. Nowcasting model is built based on the factor U-MIDAS model

Formula D(RGDP) ~ mls(D(RGDP), 1:2, 1) + mls(VN, 0:2, 1) + mls(PMI, 0:2, 1) + mls(PC1,
0:7, 3) + mls(PC2, 0:3, 3)

	Estimate	Std. Error	Pr(>|t|)		Estimate	Std. Error	Pr(>|t|)
(Intercept)	-2.54E-03	4.58E-14	1.15E-11***	PC13	4.33E-02	5.81E-14	8.54E-13***
RGDP1	9.95E-01	5.14E-12	13.29E-12***	PC14	2.19E-03	3.13E-14	9.08E-12***
RGDP2	-5.20E-01	3.47E-12	4.25E-12***	PC15	2.82E-02	2.55E-13	5.77E-12***
VN1	1.31E-01	1.98E-13	9.61E-13***	PC16	-6.33E-02	2.04E-13	2.05E-12***
VN2	-2.19E-01	1.78E-13	5.18E-13***	PC17	-6.47E-03	1.45E-14	1.42E-12***
VN3	1.85E-01	3.23E-13	1.11E-12***	PC18	2.50E-02	6.01E-14	1.53E-12***
PMI1	-1.09E-01	3.96E-13	2.32E-12***	PC21	-2.71E-02	3.09E-14	7.27E-13***
PMI2	-1.25E-01	4.96E-13	2.53E-12***	PC22	2.27E-02	1.76E-14	4.94E-13***
PMI3	-2.74E-01	5.64E-13	1.31E-12***	PC23	-9.23E-03	2.69E-14	1.86E-12***
PC11	2.35E-02	1.25E-13	3.39E-12***	PC24	9.28E-03	2.16E-14	1.48E-12***
PC12	-5.50E-02	4.45E-14	5.16E-13***				

where Z_i (i=1, 2, ...) is the variable Z lagged i-1 months. MLS() is the function in the 'MIDASu'
package in R.CRAN that presents data to perform the MIDAS regression.
　Significant codes: 0.0001: '****'; 0.001 '***'; 0.01 '**'; 0.05 '*'; 0.1: '.'; 1: ' '.
　- RMSE (dynamic forecast): 0.00204 on 1 degree of freedom;
　- The maximum lag for all factors at monthly frequency: 08;
　- The maximum lag for all variables at quarterly frequency: 03.

Since the RMSE of the nowcasting model of the RGDP indicator built based on the
factor bridge equation model is smaller than that based on the factor U-MIDAS model,
the factor bridge equation model is selected to perform the following stages.

Table 3 below shows the number of chosen factors using the KTPCA method based
on an RMSE-best model with kernels that are not the inner product, the cumulative
eigenvalue percentage, and the RMSE of the RGDP indicator nowcasting model on the
variables VN, PMI, and these two factors, where the nowcasting model is built based
on the factor bridge equation model.

Table 3. Results of variable dimension reduction and nowcasting model building of the proposed method

Kernels $\kappa(x_i, x_j) =$	Number of chosen factors	Cumul. Eigen. percentage (%)	RMSE of the model	Maximum lag
$\langle x_i, x_j \rangle$ - (PCA)	2	77.496	0.000796	3
$\left(\langle x_i, x_j \rangle + 0.5\right)^2$	1	99.21	0.003023	3
$\left(\langle x_i, x_j \rangle + 0.5\right)^3$	1	99.97	0.003019	3
$\exp(-\frac{\|x_i-x_j\|^2}{2.e^{1.464}})$	1	80.60	0.003084	3
$\exp(-\frac{\|x_i-x_j\|^2}{2.e^{0.5}})$	6	76.14	0.002525	3
$\exp(-\frac{\|x_i-x_j\|^2}{2.e^{2.5}})$	1	90.47	0.003155	3

Table 3 shows that the inner product of two vectors is the most suitable kernel among the reported kernels to perform the variable dimension reduction by the KTPCA method based on an RMSE-best model. The model having the smallest RMSE among the built nowcasting models is the model in Table 2a. Figure 3 shows graphs of the actual RGDP, the fitted RGDP produced by that model, and its RMSE, where the RMSE graph lying between the two parallel lines implies that the model's forecast error (or RMSE) is relatively small and it can be considered to be approximately equal to zero and no observation is abnormal.

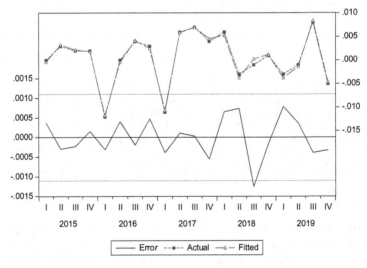

Fig. 3. Graphs of the Actual and Fitted RGDP indicator and RMSE of the nowcasting model

5   Update Forecasts of the RGDP Indicator in Real-Time Data Flow

One of the essential applications of nowcasting models is update forecasts by real-time data flows.

5.1   Handling Ragged-Edge Data in Nowcasting Model of Macroeconomic Indicators

Ragged-edge data often happens in mixed-frequency datasets due to missing values at the end samples for some predictors. There are three different methods to tackle it [6–8, 37]. The work [6] proposed realigning each time series in the sample to obtain a balanced dataset and then estimate the factors with the dynamic PCA method. The work [7] handled missing values in datasets using the EM algorithm and the traditional dynamic PCA method. The work [8] used the factor estimation approach based on a complete parametric representation of the factor model in state-space form. Developed on the idea of [6], the article proposes the realign of ragged-edge data to obtain a balanced dataset at the end samples of the predictors as follows:

Assuming T, t are the current quarter and month respectively, then t = 3T-2, 3T-1, or 3T (T = 1, 2, 3, 4). Our purpose is to forecast the quarter RGDP for the current and next quarter (here if (T + 1) > 5, then T = (T + 1) mode 4) whenever new data of the predictors is released on a date in month t.

- For predictors at the monthly frequency in Table 1, the new data updates are only made on the end days of the month (or the beginning days of the following month). Therefore, the forecasted value of these predictors in the following month (in this month) is the current value of the predictors in that month.
- Assume that X_t^M is a monthly frequency predictor aggregated from a daily frequency predictor X^D in a month t (for example, stock indices, exchange rates,...., in Table 1 are such predictors), X_t^{fM} is the forecast of the X^{TM} done on the most recent date. On a day in the month t, data of the X^D predictor is released, and its aggregated predictor at the monthly frequency X_t^M is updated as follows:

$$X_t^M = \frac{(X_{D,1} + X_{D,2} + \ldots + X_{D,m}) + (N - m) * X_t^{fM}}{N} \tag{8}$$

where $X_{D,i}$ is the value of the X^D on the ith working day from the first working day in the month t, N is the number of working days in the month t, $m > 0$ is the number of working days from the first working day in the month t to the date on which the data of variable X^D is released.

- Assume that X_T^Q is a quarterly frequency predictor aggregated from the monthly frequency predictor X_t^M in the month t of the quarter T, here X_t^M is the monthly frequency predictor aggregated from the daily frequency predictor X^D (in Table 1, it is the VN predictor); X_T^{fQ} is the forecast of the X_T^{fM} done on the most recent date. On a day in the month t, the data of X^D predictor is released, and the following formula updates its aggregated predictor X_T^Q at the quarterly frequency:

$$X_T^Q = \begin{cases} \frac{X_t^M + X_{t+1}^{fM} + X_{t+2}^{fM}}{3} \ if \ t = 3T - 2 \\ \frac{X_{t-1}^M + X_t^M + X_{t+1}^{fM}}{3} \ if \ t = 3T - 1 \\ \frac{X_{t-2}^M + X_{t-1}^M + X_t^M}{3} \ if \ t = 3T \end{cases} \tag{9}$$

where X_{t-k}^M is the X_t^{TM} predictor lagged k months, and X_{t+k}^{fM} is the out-of-sample k months forecast of the X_t^{TM} predictor from the date on which data of the X^D is released. The forecast of X_T^Q in the next quarter is based on the updated X_T^Q using an auxiliary model.

- Assume that Z_t^M is a monthly frequency predictor aggregated from a weekly frequency predictor Z^W in a month t (in Table 1, it is a lending rate or an interest rate). On a day in the month t, the data of Z^W predictor is released, and its aggregated predictor Z_t^M at the monthly frequency is updated by

$$Z_t^M = [m * Z_{t-1}^M + \sum_{i=1}^{P} k_i * Z_{t,i}^W + (N - m - \sum_{i=1}^{P} k_i) * Z_t^{fM}]/N \tag{10}$$

here $Z_{t,i}^W$ is the value of the weekly frequency predictor Z^W at the ith week in the month t; $1 \leq p \leq 5$; k_i is the number of days in the month t where the value of this variable is $Z_{t,i}^W$; $m > 0$ is the number of days from the first day in the month t to the date on which data of the weekly frequency predictor Z^W is firstly released.

In other words, on a day in a month t of the current quarter, if the data of predictors at the daily and weekly frequency is released, the value of predictors at the monthly frequency aggregated from the preditors at a higher frequency is updated and used to forecast values of the aggregated predictors in the following months. As for predictors at the monthly frequency, the value of the predictors at the month t is its forecasted value implemented at the latest update.

5.2   Sample Extension

It is necessary to develop a method of variable dimension reduction of sample extensions to reuse the variable dimension reduction previous results whenever new observations of the predictors appear. The method proposed in this article is similar to the method in [31].

Assuming $\mathbf{A}_{h \times N}$ is an h × N matrix of h new observations of N predictors, then the corresponding extension of p factors $\mathbf{PC}_{h \times p}$ of the h observations is determined by the following formula:

$$\mathbf{PC}_{h \times p} = (\mathbf{A}_{h \times N} - \overline{\mathbf{X}}) \times \mathbf{E}_{N \times p} \tag{11}$$

here $\overline{\mathbf{X}} = [X_1 - \mu_1, X_2 - \mu_2, \ldots, X_N - \mu_N]$, where $X_i = (x_{ij}), i = 1, \ldots, m, j = 1, \ldots, N$ and $\mu_i = (\frac{1}{m+h} \sum\limits_{j=1}^{m+h} x_{ij})_{1Xm}$, for every i = 1, ... , N. $\mathbf{E}_{Nxp}$ is the matrix of the first p eigenvectors of the kernel matrix of the original input dataset.

5.3   Update Forecast of the RGDP Indicator

Figure 4 below illustrates the updating of Vietnam's RGDP forecasts according to data release dates of the input predictors in Table 1.

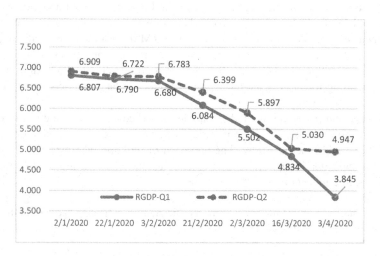

Fig. 4. Real-time forecasts of Vietnam's quarter RDGP

As known, 2019 is the year that Vietnam's economy has developed very well compared to the previous ten years: high economic growth, low inflation, high industrial production indexes, low unemployment rate, high total import and export, and the highest trade surplus ever, public debt decreased, Vietnam's stock index VNindex increased sharply. Vietnam's RGDP in the 1–4 quarters of 2019 was 6.8%, 6.8%, 7.31%, and 6.97%, respectively. The economy continued to develop exceptionally well until the end of January 2020, and since then, the implementation of many measures to combat the Covid-19 epidemic has strongly affected economic activities. The actual RGDP in the first quarter, 2020 is only **3.82%** (released by the GSO on March 30, 2020), much lower than in previous years.

Updating the real-time forecasts of RGDP in Fig. 3 shows that on January 2, 2020, Vietnam's quarter RGDP in the first and second quarters were forecasted to be relatively high, almost equal to the RGDP of the first and second quarters of the year 2019, respectively. However, from the beginning of February 2020, Vietnam's quarter RGDP in the first and second quarters decreased rapidly according to the extent of social distancing measures of Covid-19 epidemic prevention.

6   Conclusions

This article built the nowcasting model of the RGDP indicator based on a large number of potential predictors at many different frequencies using the most suitable model in the two models to be the factor bridge equation model and the factor U-MIDAS. Here, the factors are extracted in the set of original predictors using a combination of two-dimensional reduction techniques: feature selection using the Pearson correlation coefficient measure and feature learning technique using the KTPCA method based on an RMSE-best model. The built model is used to update forecasts of the RGDP indicator in real-time data follow.

The feature selection technique based on the Pearson correlation coefficient measure is used to eliminate noisy and redundant information, while the feature learning technique using the KTPCA method based on an RMSE-best model is used to extract a few new factors, but it still retains as much crucial information as possible of the dataset of original predictors. According to the proposed variable dimension reduction method, building forecasting and nowcasting models on large datasets are not divided into two phases but aggregated in one.

The paper proposes the method to handle ragged-edge data and the reinforcement learning method to reuse the previous variable dimension reduction results for updating forecasts in a real-time data flow. The approach of building the nowcasting model of the RGDP indicator in this article can be used to build nowcasting models for other macroeconomic indicators.

Funding. This research did not receive any specific grant from funding agencies in the public, commercial, or not-for-profit sectors.

References

1. Giannone, D., Reichlin, L., Small, D.H.: Nowcasting: the real-time informational content of macroeconomic data. J. Monet. Econ. **55**(4), 665–676 (2008)
2. Kapetanios, G., Papailias, F., et al.: Big Data & Macroeconomic Nowcasting: Methodological Review. Economic Statistics Centre of Excellence, National Institute of Economic and Social Research (2018)
3. Bok, B., Caratelli, D., Giannone, D., Sbordone, A.M., Tambalotti, A.: Macroeconomic nowcasting and forecasting with big data. Ann. Rev. Econ. **10**, 615–643 (2018)
4. Baldacci, E., et al.: Big Data and Macroeconomic Nowcasting: From Data Access to Modelling. Luxembourg: Eurostat. http://dx.doi.org/10.2785/360587 (2016)
5. Doornik, J.A., Hendry, D.F.: Statistical model selection with 'big data.' Cogent Econ. Finance **3**(1), 1045216 (2015)
6. Altissimo, F., Cristadoro, R., Forni, M., Lippi, M., Veronese, G.: New Euro coin: tracking economic growth in real-time. Rev. Econ. Stat. **92**(4), 1024–1034 (2010)
7. Stock, J.H., Watson, M.W.: Forecasting using principal components from a large number of predictors. J. Am. Stat. Assoc. **97**(460), 1167–1179 (2002)
8. Doz, C., Giannone, D., Reichlin, L.: A two-step estimator for large approximate dynamic factor models based on Kalman filtering. J. Econ. **164**(1), 188–205 (2011)

9. Kitchin, R., McArdle, G.: What makes big data, big data? exploring the ontological characteristics of 26 datasets. Big Data Soc. **3**(1), 2053951716631130 (2016)
10. Giannone, D., Reichlin, L., Small, D.H.: Nowcasting RGDP and inflation: the real-time informational content of macroeconomic data releases. ECB Working Article no. 633 (2006). 51p. http://hdl.handle.net/10419/153067
11. Panagiotelis, A., Athanasopoulos, G., Hyndman, J.H., Jiang, B., Vahid, F.: Macroeconomic forecasting for australia using a large number of predictors. Int. J. Forecast. **35**(2), 616–633 (2019)
12. Kim, H.H., Swanson, N.R.: Methods for Pastcasting, Nowcasting, and Forecasting Using Factor-MIDAS with an Application to Real-Time Korean GDP. Mimeo, Rutgers University (2015). 51p.
13. Kim, H.H., Swanson, N.R.: Mining big data using parsimonious factor, machine learning, variable selection, and shrinkage methods. Int. J. Forecast. **34**(2), 339–354 (2018)
14. Chikamatsu, K., Hirakata, N., Kido, Y., Otaka, K., et al.: Nowcasting Japanese GDPs. Bank of Japan (2018)
15. Bragoli, D.: Now-casting the Japanese economy. Int. J. Forecast. **33**(2), 390–402 (2017)
16. Castle, J.L., Hendry, D.F., Kitov, O.I.: Forecasting and Nowcasting Macroeconomic Variables: A Methodological Overview. Discussion Paper No. 674. University of Oxford (2013). 73p. ISSN 1471-0498
17. Foroni, C., Marcellino, M.: A comparison of mixed frequency approaches for nowcasting Euro area macroeconomic aggregates. Int. J. Forecast. **30**(3), 554–568 (2014)
18. Foroni, C., Marcellino, M.G.: A Survey of Econometric Methods for Mixed-Frequency Data (2013). Available at SSRN 2268912
19. Guresen, E., Kayakutlu, G., Daim, T.U.: Using artificial neural network models in stock market index prediction. Expert Syst. Appl. **38**(8), 10389–10397 (2011)
20. Kara, Y., Boyacioglu, M.A., Baykan, Ö.K.: Predicting the direction of stock price index movement using artificial neural networks and support vector machines: the sample of the Istanbul stock exchange. Expert Syst. Appl. **38**(5), 5311–5319 (2011)
21. Wang, J., Wang, J.: Forecasting stock market indexes using principal component analysis and stochastic time-effective neural networks. Neurocomputing **156**, 68–78 (2015)
22. Hoseinzade, E., Haratizadeh, S.: CNNpred: CNN-based stock market prediction using a diverse set of variables. Expert Syst. Appl. **129**, 273–285 (2019)
23. Lauzon, F.Q.: An introduction to deep learning. In: 11th International Conference on Information Science, Signal Processing and Their Applications (ISSPA), pp. 1438–1439 (2012)
24. Bańbura, M., Rünstler, G.: A look into the factor model black box: publication lags and the role of hard and soft data in forecasting GDP. Int. J. Forecast. **27**(2), 333–346 (2011)
25. Urasawa, S.: Real-time GDP forecasting for Japan: a dynamic factor model approach. J. Jpn. Int. Econ. **34**, 116–134 (2014)
26. Baffigi, A., Golinelli, R., Parigi, G.: Bridge models to forecast the Euro area GDP. Int. J. Forecast. **20**(3), 447–460 (2004)
27. Ghysels, E., Santa-Clara, P., Valkanov, R.: The MIDAS Touch: Mixed Data Sampling Regression Models (2004). https://escholarship.org/uc/item/9mf223rs
28. Bai, J., Ghysels, E., Wright, J.H.: State space models and MIDAS regressions. Econ. Rev. **32**(7), 779–813 (2013)
29. Ankargren, S., Lindholm, U.: Nowcasting Swedish RGDP Growth. Working Article 154, Published by the National Institute of Economic Research (NIER) (2021). ISSN 1100-7818, 33p.
30. Shlens, J.: A tutorial on principal component analysis. ArXiv Preprint ArXiv:1404.1100 (2014)

31. Van Der Maaten, L., Postma, E.: Dimensionality reduction: a comparative review. J. Mach. Learn. Res. **10**, 66–71 (2009)
32. Greene, W.H.: Econometric Analysis. Prentice-Hall (2002). ISBN 0-13-066189-9
33. Ghysels, E., Kvedaras, V., Zemlys, V.: Mixed frequency data sampling regression models: the R package MIDAS. J. Stat. Softw. **72**(1), 1–35 (2016)
34. Kim, K.I., Franz, M.O., Scholkopf, B.: Iterative Kernel principal component analysis for image modeling. IEEE Trans. Pattern Anal. Mach. Intell. **27**(9), 1351–1366 (2005)
35. Schölkopf, B., Smola, A.: A short introduction to learning with Kernels. In: Mendelson, S., Smola, A.J. (eds.) Advanced Lectures on Machine Learning. LNCS (LNAI), vol. 2600, pp. 41–64. Springer, Heidelberg (2003). https://doi.org/10.1007/3-540-36434-X_2
36. Ma, X., Zabaras, N.: Kernel principal component analysis for stochastic input model generation. J. Comput. Phys. **230**(19), 7311–7331 (2011)
37. Marcellino, M., Schumacher, C.: Factor MIDAS for nowcasting and forecasting with ragged-edge data: a model comparison for German GDP. Oxford Bull. Econ. Stat. **72**(4), 518–550 (2010)

Predicting Vietnamese Stock Market Using the Variants of LSTM Architecture

Cong-Doan Truong[1(✉)], Duc-Quynh Tran[1], Van-Dinh Nguyen[1], Huu-Tam Tran[2], and Tien-Duy Hoang[3]

[1] International School, Vietnam National University, Hanoi, Vietnam
{doantc, quynhtd, dinhnv}@isvnu.vn
[2] Faculty of Information Technology, Hanoi University, Hanoi, Vietnam
tam_fit@hanu.edu.vn
[3] FPT University, Hanoi, Vietnam
duy18mse13013@fsb.edu.vn

Abstract. Recently, the problem of stock market prediction has attracted a lot of attention. Many studies have been proposed to apply to the problem of stock market prediction. However, achieving good results in prediction is still a challenge in research and there are very few studies applied to Vietnamese stock market data. Therefore, it is necessary to improve or introduce new forms of prediction. Specifically, we have focused on the stock prediction problem for the Vietnamese market in the short and long term. Long short-term memory (LSTM) based on deep learning model has been applied to big data problem such as VN-INDEX. We compared the prediction results of the variants of the LSTM model with each other. The results obtained are very interesting that the Bidirectional LSTM architecture gives good results in short- and long-term prediction for the Vietnamese stock market. In conclusion, the LSTM architecture is very suitable for the stock prediction problem in the long- and short- term.

Keywords: LSTM · Bidirectional LSTM · VN-INDEX · Stock market prediction

1 Introduction

In fact, there is a lot of research related to stock predictions from many fields such as economics and computer science [1–3]. In computer science, scientists often use historical data such as opening prices, closing prices, highest price, lowest price, and trading volumes to predict the future prices of stocks [4]. There are many methods to forecast the stock market. Traditional methods are ARIMA (Autoregressive Integrated Moving Average) [5], ARMA (Autoregressive Moving Aver-age) [6], AR (Autoregressive) [7]. The advantage of these methods is that they are well suited to time series data such as stock markets. However, these methods give poorly results of prediction for big data and have not learned all the features of the data through the training process compared to modern methods such as deep learning [8].

© ICST Institute for Computer Sciences, Social Informatics and Telecommunications Engineering 2021
Published by Springer Nature Switzerland AG 2021. All Rights Reserved
P. Cong Vinh and N. Huu Nhan (Eds.): ICTCC 2021, LNICST 408, pp. 129–137, 2021.
https://doi.org/10.1007/978-3-030-92942-8_11

Recently, deep learning methods have been widely used for classification and prediction problems [9]. Because of its good computational infrastructure, this method is applied to many fields such as speech recognition, financial analysis, image processing, and natural language processing [10, 11]. The deep learning methods can train big data and learn data features, so this method gives good results. In fact, the data of the stock prediction problem is big data because it is collected from many transactions over ten years at stock exchanges. Recently, some studies have been applied to the stock market prediction problem [12–14]. However, these methods are often applied to data from international exchanges such as the SP&500, Nikkei, and Shanghai Index. There are very few applied studies on the VN-INDEX dataset of the Ho Chi Minh Stock Exchange. Besides, these studies only predict in the short-term period. Actually, studies with long-term predictions are not available.

For the researches in Vietnam in recent years, there are also some studies on the prediction problem for the stock exchange [15]. In particular, there are studies that use Internet news to make predictions related to the rise and fall of the stock market [16, 17]. In addition, there are studies that evaluate the role of investor behavior in the stock market based on the relationship between investor behavior and profits [18]. The common feature of recent studies is to focus on studying each news, political, and social relationship that affects stock market prices. In addition, there are a number of studies that apply deep learning techniques to predict stock prices [19, 20]. However, these studies have not focused on the problem of long-term prediction for the increase or decrease of stock prices.

Therefore, in this paper, we focus on researching using data of VN-INDEX of Ho Chi Minh City exchange over a long period of time. We choose the closing price to characterize the predictive analysis. With a large amount of data, we applied deep learning to two basic problems: short-term and long-term prediction. The very interesting results obtained are that the Bidirectional LSTM is well suited for short-term and long-term stock prediction.

The rest of this paper is organized as follows. Section 2 describes the methodology. Section 3 presents the experiments and their results. Finally, the main findings of this paper are concluded in Sect. 4.

2   Preparations and Methodology

2.1   Datasets

In this study, we used the VN-INDEX database on securities of Ho Chi Minh City exchange. Transactions are collected from the date 31st of July, 2000 to the date 25th of March, 2021. VN-INDEX stock data has been used in many recent studies [15, 21]. However, our study collected data for the most recent new transactions compared to other studies.

2.2   Introduction to Deep Recurrent Neural

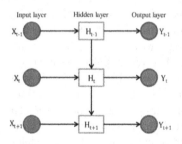

Fig. 1. A network architecture of deep recurrent neural

Deep learning is used to design computational layers that can learn from input data. Each layer in the architecture is designed to output informational features from the input. The first output of each layer is the beginning of the next layer. One of the techniques used for architecture of construction is RNN. This architecture uses input data to learn weights in the network. While the network is built from scratch with data, the special importance of the information will be stored in the hidden network states. Backpropagation is used to construct technical trainings for the network. calculations will be performed in sequence time and linked steps. Figure 1 shows the architectural benchmark of a deep RNN and the processes of prediction.

2.3   Definition of Long-Short Term Memory (LSTM)

The LSTM is proposed by Hochreiter and Schmidhuber [22], is one of the variants of RNN to solve the gradient loss problem. It can define ports, which can receive information by keeping vital information of the data. A learning process through backpropagation can estimate weights to allow data in cells to be saved or deleted. The formulas of the LSTM are as follows:

$$i_t = \sigma(W_i x_t + U_i h_{t-1} + V_i c_{t-1})$$
$$f_t = \sigma(W_f x_t + U_f h_{t-1} + V_f c_{t-1})$$
$$o_t = \sigma(W_o x_t + U_o h_{t-1} + V_o c_t)$$
$$\tilde{c}_t = tanh(W_c x_t + U_c h_{t-1})$$
$$c_t = f_t^i \odot c_{t-1} + i_t \odot \tilde{c}_t$$
$$h_t = o_t \odot + tanh(c_t)$$

where i_t indicates the input port and o_t indicates the output port. Where f, c_t, and h_t indicate forget gate, memory cell, and hidden state, respectively.

2.4 Definition of Bidirectional LSTM (BLSTM)

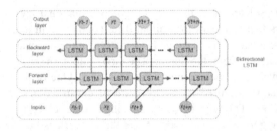

Fig. 2. Architecture of bidirectional RNN

One of the variants of RNN is Bidirectional RNN (BRNNs), this model was developed by Schuster and Paliwal [23]. BRNNs connects two hidden layers with opposite directions but same output. With This design, the output layer can get information from the backwards and from the forward states at the same time. BRNNs are proposed to enhance the processing of input data for the network. Figure 2 shows the effectiveness of BRNNs over unidirectional LSTMs in some application problems such as classification.

2.5 Formulation of Short-Term and Long-Term Prediction and Evaluation of Performance of Models

In this article, we use data of 15 days to predict the next day's price of a security and make a long-term prediction for the next 60 days. Besides, we used 3 error measures as follows:

$$MAE = \frac{1}{n} + \sum_{t=0}^{n-1} |y_t - \hat{y}_t| \qquad (1)$$

$$RMSE = \sqrt{\frac{1}{n} \sum_{t=0}^{n-1} (y_t - \hat{y}_t)^2} \qquad (2)$$

$$R^2 = 1 - \frac{\sum_{t=0}^{n-1} (y_t - \hat{y}_t)^2}{\sum_{t=0}^{n-1} (y_t - \overline{y})^2} \qquad (3)$$

Formula (1) is Mean Absolute Error (MAE), formula (2) is Root Mean Square Error (RMSE) and formula (3) is R^2. MAE and RMSE if close to zero is better for model prediction, otherwise R^2 is closer to 1 is better.

3 Results

We have conducted experiments using Python programming language and Keras open source library for deep learning with TensorFlow backend [24]. The closing price has been chosen as the only feature used to predict and evaluate the predictive capacity of the model. We have done the practice of dividing the data set into two parts. Part 1 takes up 80% of all data, and part 2 takes 20% for testing. We have also used this same data set for different algorithms LSTM, SLSTM and BLSTM. We have conducted using three error measures that are MAE, RMSE and R^2. Next, I experimented four times with different numbers of neurons increasing from 5, 10, 20, and 40 for both short-term and long-term prediction problems with all three models. The tables show the prediction results of the three models LSTM, SLSTM, and BLSTM, below.

Table 1. Short-term prediction results of LSTM model

Networks	MAE		RMSE		R square
	Testing	Training	Testing	Training	
5 neurons network	5.780	4.902	46.568	32.271	0.829
10 neurons network	4.316	3.660	26.53	20.473	0.944
20 neurons network	4.617	3.681	27.991	20.740	0.938
40 neurons network	4.38	3.594	26.000	20.333	0.946
Total average	4.773	3.959	31.772	23.454	0.914

Table 2. Short-term prediction results of SLSTM model

Networks	MAE		RMSE		R square
	Testing	Training	Testing	Training	
5 neurons network	5.730	4.872	46.287	31.963	0.833
10 neurons network	5.097	4.277	36.396	26.468	0.897
20 neurons network	5.321	4.0481	36.170	24.926	0.898
40 neurons network	4.735	3.898	30.260	23.201	0.929
Total average	5.220	4.273	37.278	26.639	0.8892

Table 3. Short-term prediction results of BLSTM model

Networks	MAE		RMSE		R square
	Testing	Training	Testing	Training	
5 neurons network	4.821	3.893	30.435	21.767	0.928
10 neurons network	4.071	3.334	23.383	17.488	0.957
20 neurons network	4.236	3.535	25.400	19.826	0.949
40 neurons network	4.180	3.532	16.848	19.683	0.977
Total average	4.327	3.573	24.016	19.691	0.952

Table 4. Long-term prediction results of LSTM model

Networks	MAE		RMSE		R square
	Testing	Training	Testing	Training	
5 neurons network	5.578	6.632	40.114	67.295	0.874
10 neurons network	4.755	6.951	28.202	70.524	0.938
20 neurons network	8.283	6.726	74.861	70.267	0.564
40 neurons network	7.009	6.886	56.665	67.661	0.750
Total average	6.406	6.798	49.96	68.936	0.781

Table 5. Long-term prediction results of SLSTM model

Networks	MAE		RMSE		R square
	Testing	Training	Testing	Training	
5 neurons network	6.543	6.955	57.997	75.069	0.738
10 neurons network	7.008	6.915	61.34	74.578	0.707
20 neurons network	6.828	6.89	58.506	74.408	0.733
40 neurons network	5.329	7.492	38.45	78.215	0.885
Total average	6.427	7.063	54.073	75.567	0.765

Table 6. Long-term prediction results of BLSTM model

Networks	MAE		RMSE		R square
	Testing	Training	Testing	Training	
5 neurons network	4.981	7.027	33.635	70.004	0.912
10 neurons network	4.757	7.164	29.799	72.274	0.930
20 neurons network	4.513	6.992	27.566	70.141	0.940
40 neurons network	6.363	6.582	47.806	67.360	0.822
Total average	5.153	6.941	34.701	69.944	0.901

Based on the experimental results, we have given in the above six tables the results. We can comment as follows: Tables 1, 2, and 3 show the results for the short-term prediction of three models: LSTM, SLSTM, and BLSTM. Tables 4, 5, and 6 are the results of long-term stock price predictions. For short-term prediction, the average result of LSTM on the test data set with error measures MAE, RMSE, and R^2 is 4,773, 31,772, and 0.914, respectively, and the average result of SLTSM on the test data set with error measures MAE, RMSE and R2 are 5,220, 37,278, and 0.8892, respectively, the average result of BLSTM is 4.237, 24,016 and 0.952, respectively. Based on this result, we can see that the BLSTM model gives better results than the LSTM and SLSTM models when it comes to short-term prediction. Similarly, we evaluate the results when predicting long-term with three models LSTM, SLSTM, and BLSTM. The average result of the LSTM with the three error measures MAE, RMSE, and R^2 is 6,406, 49,960, and 0.781, respectively, and the mean result of the SLSTM with MAE, RMSE, and R^2 is 6,427, 54,073, and 0.765, respectively. The results of the BLSTM model with three error measures MAE, RMSE, and R^2 are 5.153, 34,701, and 0.901, respectively. Based on this result, we can say that the BLSTM model gives the best results on the test set with error measures MAE, RMSE, and R^2 for both short-term and long-term prediction problems for the VN-INDEX.

We conduct one more experiment, we proceed to select the best model of LSTM, SLSTM, and BLSTM. The results for the short-term prediction problem are shown in Table 7 below. Based on the results of Table 7, we can see that the BLSTM model gives good predictive results on both the training and testing datasets. Similarly, the best models for long-term prediction are shown in Table 8. The results show that the BLSTM model gives the best prediction results for the long-term prediction problem on the testing dataset. In addition, we simulated the results of the models in Figs. 3 and 4.

Table 7. Results of comparing the best models for short-term prediction problem

	Training			Testing		
	MAE	RMSE	R^2	MAE	RMSE	R^2
LSTM	3.594	20.333	0.946	4.380	26.000	0.946
SLSTM	3.898	23.201	0.929	4.735	30.260	0.929
BLSTM	3.532	19.683	0.977	4.180	16.848	0.977

Table 8. Long-term prediction results of the best models

	Training			Testing		
	MAE	RMSE	R^2	MAE	RMSE	R^2
LSTM	6.886	67.661	0.75	7.009	56.665	0.75
SLSTM	7.492	78.215	0.885	5.329	38.45	0.885
BLSTM	6.992	70.141	0.94	4.513	27.566	0.94

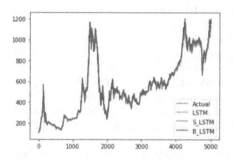

Fig. 3. Short-term prediction results using LSTM, SLSTM, and BLSTM

Fig. 4. Long-term prediction results of LSTM, SLSTM, and BLSTM models

4 Conclusion

In this paper, we have built a stock prediction problem. The selected data is VN-INDEX collected from the Ho Chi Minh City exchange from 31st of July, 2000 to 25th of March 2021. We have performed predictions for two problems which are short-term and long-term prediction problems. Three prediction models have been selected, namely LSTM, SLSTM, and BLSTM. The results show that the BLSTM model performed better than other models for this time -series data problem. In other words, the BLSTM model could be a favorable method to choose for predicting the Vietnamese stock market both in short and long-term periods.

Acknowledgement. This research is funded by International School, Vietnam National University, Hanoi (VNU-IS) under project number CS.2021-02.

References

1. Nti, I.K., Adekoya, A.F., Weyori, B.A.: A novel multi-source information-fusion predictive framework based on deep neural networks for accuracy enhancement in stock market prediction. J. Big Data **8**(1), 1–28 (2021). https://doi.org/10.1186/s40537-020-00400-y
2. Jiang, W.: Applications of deep learning in stock market prediction: recent progress. Expert Syst. Appl. **184**, 115537 (2021). https://doi.org/10.1016/j.eswa.2021.115537
3. Ananthi, M., Vijayakumar, K.: Stock market analysis using candlestick regression and market trend prediction (CKRM). J. Ambient Intell. Humaniz. Comput. **12**(5), 4819–4826 (2020). https://doi.org/10.1007/s12652-020-01892-5
4. Wang, X., Phua, P.K.H., Lin, W.: Stock market prediction using neural networks: does trading volume help in short-term prediction? In: Proceedings of the International Joint Conference on Neural Networks, vol. 4, pp. 2438–2442 (2003). https://doi.org/10.1109/IJCNN.2003.1223946
5. Ariyo, A.A., Adewumi, A.O., Ayo, C.K.: Stock price prediction using the ARIMA model. In: 2014 UKSim-AMSS 16th International Conference on Computer Modelling and Simulation, pp. 106–112 (2014). https://doi.org/10.1109/UKSim.2014.67
6. Anaghi, M.F., Norouzi, Y.: A model for stock price forecasting based on ARMA systems. In: 2012 2nd International Conference on Advances in Computational Tools for Engineering Applications (ACTEA), pp. 265–268 (2012). https://doi.org/10.1109/ICTEA.2012.6462880
7. Hushani, P.: Using autoregressive modelling and machine learning for stock market prediction and trading. In: Yang, X.-S., Sherratt, S., Dey, N., Joshi, A. (eds.) Third International Congress on Information and Communication Technology. AISC, vol. 797, pp. 767–774. Springer, Singapore (2019). https://doi.org/10.1007/978-981-13-1165-9_70
8. S. Siami-Namini, N., Tavakoli, A., Namin, S.: A comparison of ARIMA and LSTM in forecasting time series. In: 2018 17th IEEE International Conference on Machine Learning and Applications (ICMLA), pp. 1394–1401 (2018). https://doi.org/10.1109/ICMLA.2018.00227
9. Huang, J., Chai, J., Cho, S.: Deep learning in finance and banking: a literature review and classification. Front. Bus. Res. China **14** (2020). https://doi.org/10.1186/s11782-020-00082-6
10. Voulodimos, A., Doulamis, N., Doulamis, A., Protopapadakis, E.: Deep learning for computer vision: a brief review. Comput. Intell. Neurosci. **2018**, 7068349 (2018). https://doi.org/10.1155/2018/7068349
11. Wu, S., et al.: Deep learning in clinical natural language processing: a methodical review. J. Am. Med. Inform. Assoc. **27**, 457–470 (2020). https://doi.org/10.1093/jamia/ocz200
12. Hiransha, M., Gopalakrishnan, E.A., Menon, V.K., Soman, K.P.: NSE stock market prediction using deep-learning models. Procedia Comput. Sci. **132**, 1351–1362 (2018). https://doi.org/10.1016/j.procs.2018.05.050
13. Vargas, M.R., de Lima, B.S.L.P., Evsukoff, A.G.: deep learning for stock market prediction from financial news articles. In: 2017 IEEE International Conference on Computational Intelligence and Virtual Environments for Measurement Systems and Applications (CIVEMSA), pp. 60–65 (2017). https://doi.org/10.1109/CIVEMSA.2017.7995302
14. Nabipour, M., Nayyeri, P., Jabani, H., Mosavi, A., Salwana, E.: Deep learning for stock market prediction. Entropy **22** (2020). https://doi.org/10.3390/e22080840
15. Co, N.T., Son, H.H., Hoang, N.T., Lien, T.T.P., Ngoc, T.M.: Comparison between ARIMA and LSTM-RNN for VN-index prediction. In: Ahram, T., Karwowski, W., Vergnano, A., Leali, F., Taiar, R. (eds.) IHSI 2020. AISC, vol. 1131, pp. 1107–1112. Springer, Cham (2020). https://doi.org/10.1007/978-3-030-39512-4_168

16. Schumaker, R.P., Chen, H.: Textual analysis of stock market prediction using breaking financial news: the AZFin text system (2009)
17. Duong, D., Nguyen, T., Dang, M.: Stock market prediction using financial news articles on Ho Chi Minh stock. Exchange (2016). https://doi.org/10.1145/2857546.2857619
18. Phan, T.N.T., Bertrand, P., Phan, H.H., Vo, X.V.: The role of investor behavior in emerging stock markets: evidence from Vietnam. Quarterly Rev. Econ. Finance (2021). https://doi.org/10.1016/j.qref.2021.07.001
19. Huynh, H.D., Dang, L.M., Duong, D.: A New Model for Stock Price Movements Prediction Using Deep Neural Network (2017). https://doi.org/10.1145/3155133.3155202
20. Le, T.D.B., Ngo, M.M., Tran, L.K., Duong, V.N.: Applying LSTM to predict firm performance based on annual reports: an empirical study from the vietnam stock market. In: Ngoc Thach, N., Kreinovich, V., Trung, N.D. (eds.) Data Science for Financial Econometrics. SCI, vol. 898, pp. 613–622. Springer, Cham (2021). https://doi.org/10.1007/978-3-030-48853-6_41
21. Lien Minh, D., Sadeghi-Niaraki, A., Huy, H.D., Min, K., Moon, H.: Deep learning approach for short-term stock trends prediction based on two-stream gated recurrent unit network. IEEE Access 6, 55392–55404 (2018). https://doi.org/10.1109/ACCESS.2018.2868970
22. Hochreiter, S., Schmidhuber, J.: Long short-term memory. Neural Comput. 9, 1735–1780 (1997). https://doi.org/10.1162/neco.1997.9.8.1735
23. Graves, A., Schmidhuber, J.: Framewise phoneme classification with bidirectional LSTM networks. In: Proceedings. 2005 IEEE International Joint Conference on Neural Networks 2005, vol. 4, pp. 2047–2052 (2005). https://doi.org/10.1109/IJCNN.2005.1556215
24. Abadi, M., et al.: TensorFlow: a system for large-scale machine learning. In: 12th USENIX Symposium on Operating Systems Design and Implementation (OSDI 2016), pp. 265–283. USENIX Association, Savannah (2016)

Dimensionality Reduction Performance of Sparse PCA Methods

Thanh Do Van[(✉)]

Faculty of Information Technology, Nguyen Tat Thanh University,
Ho Chi Minh City, Vietnam
dvthanh@ntt.edu.vn

Abstract. Today, a hot topic is building forecasting models on large data sets of economic-financial time-series predictors using dimensionality reduction methods. The forecasting models built based on the dynamic factor model in which the factors extracted by the principal component analysis (PCA for short) method or sparse PCA (SPCA) method are superior to other benchmark models in terms of the forecast accuracy of models. Many pieces of literature have considered that the dimensionality reduction performance of the SPCA method seems to be higher than that of the PCA method. However, there have been no studies comparing the dimensionality reduction performance of those two methods to date.

The purpose of this article is to overcome that inadequacy by experimentally evaluating the dimensionality reduction performance of the two methods mentioned above on ten real-world data sets. Difference from previous beliefs, the experimental results show that the dimensionality reduction performance of the PCA and SPCA methods is competitive.

Keywords: Dimensionality reduction · PCA · Sparse PCA

1 Introduction and Contributions of the Article

The two most prominent approaches used to make forecasts or classifications on large data sets are dataset size reduction and neutron network deep learning. According to [1], neutron network deep learning is only suitable for data sets with a large number of observations, but the number of variables is not too large. The forecasts or classifications on data sets with a large number of predictors or both the number of predictors and the number of observations to be large, first, need to use some techniques to reduce the dimensionality of these data sets, in which the most important is to reduce the number of predictors (or the variable dimension). In this approach, forecasting a target variable on large data sets of time series predictors consists of 2 phases.

The first phase reduces the dimensionality of an original data set by transforming this data set in a high dimensional space to a new data set in a much lower-dimensional space, and phase 2 is to perform a forecast algorithm on the obtained new data set [2]. To reduce the dimensionality of a large data set, one often uses both two techniques of feature selection and feature learning. The feature selection techniques are often used to select highly relevant predictors and nonredundant for forecasting or classifying the

P. Cong Vinh and N. Huu Nhan (Eds.): ICTCC 2021, LNICST 408, pp. 138–148, 2021.
https://doi.org/10.1007/978-3-030-92942-8_12

original data sets [3, 4]. In essence, it is a way to deal with noisy and redundant information in data sets of original predictors. In practical application, the most commonly used feature selection technique is the feature filtering technique using the Pearson correlation coefficient measure or the entropy-based mutual information measure depending on the target variable receiving numerical values or categorical values. The data set of the selected valuable predictors is called the input data set of the predictors. This set is generally still large, and it is necessary to use feature learning techniques to reduce its dimensionality. In socio-economics, the most commonly used feature learning method to reduce the dimensionality of large time-series data sets is the principal component analysis (PCA) method.

The PCA method is an unsupervised learning one. Each principal component is a linear projection of the mean-centered input data set onto an eigenvector of the input data set's covariance matrix [5]. The variance of each principal component is the eigenvalue of a corresponding eigenvector associated with this principal component. Assuming that the principal components of the input data set are sorted in descending order of their respective eigenvalues, then the cumulative variance percentage on the eigenvalue overall of the first k principal components is the percentage of information in the original data set captured by the first k principal components [5]. With a pre-defined cumulative variance percentage threshold (in the range (70%, 90%)), suppose k is the smallest integer such that the cumulative variance percentage of the first k principal components is greater than this threshold. Then, these k principal components are selected to replace the original predictors in the target variable forecasting exercises according to the original predictors or in the classification exercises of these data sets.

Currently, the building of the target variable forecasting model according to a large set of economic-financial time-series predictors is mainly based on the dynamic factor model where the factors are extracted from the input data set of the predictors using the PCA method [1]. Many empirical studies have shown that the forecast accuracy of models built under such an approach is superior to other benchmark models [6].

Zou et al. [7] argued that since each principal component is a linear combination of all the predictors of the mean-centered input data set and their loadings are often nonzero, it has may have difficulty interpreting the derived principal components. To overcome this limitation, Zou et al. [7] proposed the sparse principal component analysis (SPCA) method, where each principal component is a linear combination of several predictors. According to the preliminary assessment of the authors of this study, the forecast accuracy of forecasting models built based on the dynamic factor model where the factors extracted using the SPCA method can be improved over that the PCA method. Many other researchers also agreed with the above assessment [1, 6, 8]. However, systematic studies have not compared the performance of the dimensionality reduction of the PCA and SPCA methods to confirm the above assessment. The purpose of this article is to experimentally compare the dimensionality reduction performance of the PCA method and the SPCA methods family, including the SPCA, randomized SPCA (RSPCA for short), and Robust SPCA (ROBSPCA) methods. Here the dimensionality reduction performance of a method is measured by the root mean squared forecast error (RMSE for short) of a forecast model built based on the dynamic factor model where the factors are extracted by this dimensionality reduction method.

The article is structured as follows; next, Sect. 2 introduces some preliminaries for the following sections, Sect. 3 presents experimental data sets and methods, Sect. 4 introduces experimental results and evaluations, and the last Sect. 5 presents a few conclusions.

2   Preliminaries

2.1   PCA, SPCA, RSPCA, and ROBSPCA Methods

The PCA Method

Assuming that $\mathbf{X}$ is an input data set. It is a N × m matrix and is expressed as follows $\mathbf{X} = [X_1, X_2, \ldots, X_m]$, here $X_i = (x_{i_1}, x_{i_2}, \ldots, x_{i_N})^{\mathrm{T}} \in \mathbb{R}^{\mathbf{N}}$. N, m, and X_i are, respectively, the observation dimension, the variable dimension, and an original variable of $\mathbf{X}$. Denote $Y_{\mathrm{h}} = (y_{\mathrm{h}_1}, y_{\mathrm{h}_2}, \ldots, y_{h_N})^{\mathrm{T}} \in \mathbb{R}^{\mathbf{N}}$, where h = 1, 2,..., p and $\mathbf{Y} = [\, Y_1, Y_2, \ldots, Y_{\mathrm{p}}]$.

The variable dimension reduction is a mapping $R : \mathbb{R}^m \to \mathbb{R}^p$ $(x_{1_j}, x_{2_j}, \ldots, x_{m_j}) \mapsto (y_{1_j}, y_{2_j}, \ldots, y_{p_j})$, s.t. p ≪ m and the data set $[Y_1, Y_2, \ldots, Y_{\mathrm{p}}] = R([X_1, X_2, \ldots, X_m])$ captures important information of $\mathbf{X}$ as much as possible. As known, when data points $(x_{1_j}, x_{2_j}, \ldots, x_{m_j})$ approximate a hyperplane, the most efficient method for reducing the variable dimension of an $\mathbf{X}$ data set is the PCA [5, 9].

PCA is the unsupervised and typical feature learning method for variable dimension reduction. This method allows transforming high-dimensional data sets into a lower-dimensional subspace while retaining the maximum variance and covariance structure of data sets [5, 10, 11]. The computation of eigenvalues, eigenvectors, and their respective PCs is performed on the covariance matrix of data sets.

Assume that $\mathbf{X} = [X_1, X_2, \ldots, X_m] \in \mathbb{R}^{\mathbf{m} \times \mathbf{N}}$, here $X_i = (x_{i_1}, x_{i_2}, \ldots, x_{i_N})^{\mathrm{T}} \in \mathbb{R}^{\mathbf{N}}$, is a mean-centered matrix, i.e., $\sum_{j=1}^{N} x_{ij} = 0$, for every i = 1,..., m; $\mathbf{R}$ is the covariance matrix of $\mathbf{X}$; $\mathbf{A}$ is the matrix of the $\mathbf{R}$'s eigenvectors, and $\mathbf{F} = [PC_1, PC_2, \ldots, PC_N] \in \mathbb{R}^{\mathbf{m} \times \mathbf{N}}$ is the matrix of the principal components of $\mathbf{X}$. Then [5]:

$$\mathbf{F} = \mathbf{XA} \tag{1}$$

The column vectors in $\mathbf{A}$ are loadings vectors. With the A loadings matrix found by the tool of linear algebra as above, the principal components sequentially capture the maximum variability in the input data set. It is also possible to convert finding the $\mathbf{A}$ matrix to a problem of maximum variance. Moreover, the problem of determining the principal components by the PCA method is also found by solving the following problem using variable projection as an optimization strategy [12]:

$$\underset{\mathbf{A}}{\mathrm{minimize}} f(\mathbf{A}) = \tfrac{1}{2} \| \mathbf{X} - \mathbf{XAA}^{\mathrm{T}} \|_{\mathrm{F}}^2 \tag{2}$$
$$\text{subject to } \mathbf{AA}^{\mathrm{T}} = \mathbf{I}$$

where $\|\cdot\|_F$ is the Frobenius norm of the matrix.

The speed of finding the principal components by finding the solution of the optimal problem (2) is faster than that by the linear algebra approach.

The SPCA and RSPCA Methods

The SPCA method combines sparse regression and standard *PCs* to find a set of sparse loading vectors, i.e., loading vectors with only a few nonzero values [7]. The SPCA method proposed in [12] using variable projection as an optimization strategy is as follows:

$$\underset{\mathbf{A,B}}{\text{minimize}} f(\mathbf{A}, \mathbf{B}) = \tfrac{1}{2} \left\| \mathbf{X} - \mathbf{XBA}^{\mathsf{T}} \right\|_{\mathrm{F}}^{2} + \psi(\mathbf{B}) \tag{3}$$
$$\text{subject to } \mathbf{AA}^{\mathsf{T}} = \mathbf{I}$$

A and **B** are both $N \times N$ square matrices where **B** is a sparse weight matrix, and **A** is an orthonormal matrix, ψ denotes a sparsity inducing regularizers such as the LASSO (l_1 norm), or the elastic net. Then the set of factors is formed as $\mathbf{F} = \mathbf{XB}$.

To speed up the computations, one used matrix approximations to factorize a given matrix into a product of smaller (low-rank) matrices. Erichson et al. [12] formed a low-dimensional sketch of the data, which aims to capture the essential information of the original data set. Then the RSPCA method using variable projection as an optimization strategy can be reformulated as:

$$\underset{\mathbf{A,B}}{\text{minimize}} f(\mathbf{A}, \mathbf{B}) = \tfrac{1}{2} \left\| \hat{\mathbf{X}} - \hat{\mathbf{X}} \mathbf{BA}^{\mathsf{T}} \right\|_{\mathrm{F}}^{2} + \psi(\mathbf{B}) \tag{4}$$
$$\text{subject to } \mathbf{AA}^{\mathsf{T}} = \mathbf{I}$$

where $\hat{\mathbf{X}} \in \mathbb{R}^{\mathbf{h} \times \mathbf{N}}$ denotes the sketch of $\mathbf{X} \in \mathbb{R}^{\mathbf{m} \times \mathbf{N}}$, and the h dimensions are chosen slightly larger than the target-rank k [12].

The ROSPCA Method

To overcome the challenge that data is grossly corrupted due to measurement errors or other effects in many real-world situations, one often uses the decomposition of a data matrix into its sparse and low-rank components. Erichson et al. [12] proposed another formulation of the ROBSPCA method using variable projection as an optimization strategy. It was also formulated based on separating a data matrix into a low-rank model and a sparse model and introducing prior information like sparsity promoting regularizers. More concretely, the ROBSPCA method using variable projection is as follows:

$$\underset{\mathbf{A,B}}{\text{minimize}} f(\mathbf{A}, \mathbf{B}) = \tfrac{1}{2} \left\| \mathbf{X} - \mathbf{XBA}^{\mathsf{T}} \right\|_{\mathrm{F}}^{2} + \psi(\mathbf{B}) + \gamma \|S\|_{1} \tag{5}$$
$$\text{subject to } \mathbf{AA}^{\mathsf{T}} = \mathbf{I}$$

where matrix **S** captures grossly corrupted outliers in the original data set.

2.2 Dynamic Factor Model

Assume $Y_t, X_{i,t}$ ($i = 1, 2,..., N$) are stationary time-series [13]. If the frequency of these variables is the same, then the factor bridge equation model and the factor mixed data sampling (MIDAS) model [1, 8, 14, 15] are the autoregressive distributed lag model. This model takes the following form [13]:

$$Y_t = \sum_{k=1}^{S} b_k Y_{t-k} + \sum_{i=1}^{r} \sum_{j=0}^{r_i} \beta_{i_j} X_{i,t-j} + \sum_{j=1}^{p} \sum_{h=0}^{p_j} \gamma_{j_h} F_{j,t-h} + c + u_t \quad (6)$$

Here $X_{i,t}$ is a predictor having a strong influence on the change of the variable Y_t. $F_{j,t}$ is a factor extracted from other predictors; u_t is the model's error and is assumed to be white noise; c, b_k, β_{i_j}, and γ_{j_h} are the estimated parameters; $X_{i,t-j}$ is the variable $X_{i,t}$ lagged j steps. r_i ($i = 1,...,$ r), p_j (j = 1,...,p), and s is the maximum lags of the variables $X_{i,t}$, $F_{j,t}$, and Y_t, respectively. The maximum lag can be determined using the Akaike information criterion (AIC) or the Schwarz information criterion (SIC) [16].

Assume $\hat{Y}_t$ is the fitted variable of the target variable Y_t. It is produced by the forecasting model of the target variable. The standard mean forecast error of this model (or RMSE) is determined by [13]:

$$\text{RMSE} = \sqrt{\frac{1}{T} \cdot \sum_{i=1}^{T} \left(Y_i - \hat{Y}_i\right)^2} \quad (7)$$

The smaller the RMSE, the higher the forecast accuracy of the model. RMSE is used to evaluate the performance of a dimensionality reduction method.

3 Experimental Datasets and Method

3.1 Experimental Datasets

Ten data sets in which 03 self-collected data sets and 07 data sets in [17] are used for experiments. Three self-collected data sets are used to forecast the core stock index VN30, consumer price index (CPI), and industrial production output of the Vietnam economy (VIP), and they are named VN30, CPI, and VIP. The remaining seven datasets are named Residential Buildings [18], S&P 500, DJI and Nasdaq [19], Air Quality [20], Energy household volume [21], and SuperConductivity [22]. The VN30, CPI, VIP, and S&P 500 datasets are input data sets, i.e., they do not contain redundant and noisy information for target variable forecasting. The remaining datasets are the original datasets. The S&P 500, DJI, NASDAQ, and Air Quality datasets were supplemented with missing data using the weighted moving average method. The weights depend on each data set. The Residential Building data set is preserved after removing the Zip codes attribute. The S&P 500, DJI, and NASDAQ datasets include observations from November 1, 2010, to October 26, 2017, in their respective original datasets [19], while the dataset The Air Quality data includes observations from noon on March 11, 2004, to noon on April 4, 2005 [20] in the original dataset, where data were collected every hour. The Household Energy dataset includes observations from 5:50 a.m.

on January 11, 2016, to 11:50 a.m. on May 27, 2016, in the original dataset, where data is collected every 10 min [21]. The SuperConductivity data set is the training data set in [22]. Table 1 below shows some statistical characteristics of these data sets. In this table, the number of attributes (No. of Attributes for short) is the number of predictors excluding the target variable.

Table 1. The statistical characteristics of experimental data sets

Data sets	Type of data set	Type of attribute	No. of obser.	No. of attributes	Missing data	The target variable	Freq.
VN30	Time series	Real	366	34	No	VN30 index	Daily
CPI	Time series	Real	72	102	No	CPI index	Monthly
VIP	Time series	Real	60	265	No	Production value of industries	Monthly
Residential building	Multivariate	Real	371	27[a]	No	Sales prices	
S&P500	Time series	Real	1760	52	Yes	S&P 500 index	Daily
DJI	Time series	Real	1760	81	Yes	Dow Jones index	Daily
NASDAQ	Time series	Real	1760	81	Yes	Nasdaq index	Daily
Air quality	Time series	Real	9348	12	Yes	CO of Air	Hourly
Appliances energy	Time series	Real	19704	23	No	The energy use of appliances (wh)	Every 10 min
Super conduct.	Multivariate	Real	21263	81	No	Critical temperature	

[a]Remove the column V1: zip codes

3.2 Experimental Method

In this article, the criterion for selecting the number of extracted factors is their cumulative eigenvalue percentage [23], and forecast models are built based on Eq. (6) using the OLS regression method under ideal conditions. Namely, the maximum lag of all variables in each model is precisely determined using the Akaike information criterion (AIC) [16]. Thus, the maximum lag of the principal components extracted using different variable dimension reduction methods is generally different for each data set. In addition, all variables are tested for unit roots and converted to stationary time series before performing model estimation, and in the final estimated model, all variables are highly statistically significant, at least less than 10%. All conditions for the model's estimate to be the best, linear and unbiased (referred to as BLUE) are guaranteed. That allows improving the forecast accuracy of the estimated model [16]. To conduct the experiments, the article used packages 'Sparsepca' [24], 'Kernlable' [25], and 'Caret' [26] in R.CRAN.

4   Results and Evaluations

With the cumulative eigenvalue percentage threshold of 75% for all the aforementioned variable dimension reduction methods and all the experimental data sets, The variable dimension reduction results include the smallest number of principal components with the cumulative variance percentage greater than the predefined threshold, a specific percentage of the cumulative variance, the maximum lag of the variables in each forecasting model built based on the factors extracted by methods of PCA, SPCA, RSPCA, ROBSPCA, and RMSE of these models are presented in Table 2.

Table 2. The variable dimension reduction results of the methods

	1. The VN30 dataset				2. The CPI dataset			
Methods	PCA	SPCA	RSPCA	ROBSPCA	PCA	SPCA	RSPCA	ROBSPCA
The number of factors/ PC_s	14	14	14	15	4	4	4	4
Cumul. Eigen. percentage (%)	75.83	75.8	75.8	76.9	78.72	77.8	77.8	76.7
Maximum lag of variables	5	5	5	5	6	6	6	6
RMSE	0.1895	0.1968	0.1968	0.2054	1.4836	1.0659	1.0673	1.0659
	3. The VIP dataset				4.The Residence Building			
Methods	PCA	SPCA	RSPCA	ROBSPCA	PCA	SPCA	RSPCA	ROBSPCA
The number of factors/ PC_s	4	4	4	4	1	1	1	1
Cumul. Eigen. percentage (%)	76.19	75.2	75.2	76	99.28	99	99	98.3
Maximum lag of variables	6	6	6	6	10	10	10	10
RMSE	715.96	826.28	1373.57	2642.83	1152.4	1152.5	1152.5	1151.2
	5.The S&P500 data set				6. The DJI data set			
Methods	PCA	SPCA	RSPCA	ROBSPCA	PCA	SPCA	RSPCA	ROBSPCA
The number of factors/ PC_s	1	1	1	1	1	1	1	1
Cumul. Eigen. percentage (%)	99.96	99.8	99.8	99.8	99.5	99.2	99.2	98.9
Maximum lag of variables	5	5	5	5	5	5	5	5
RMSE	161.415	161.441	161.441	161.441	91.82	309.24	309.24	309.23
	7. The NASDAQ data set				8. Air Quality data set			
Methods	PCA	SPCA	RSPCA	ROBSPCA	PCA	SPCA	RSPCA	ROBSPCA
The number of factors/ PC_s	1	1	1	1	1	1	1	1
Cumul. Eigen. percentage (%)	99.67	99.4	99.4	99.1	75.96	75.7	75.7	74.8
Maximum lag of variables	5	5	5	5	12	12	12	12
RMSE	365.97	85.47	85.47	85.46	71.459	71.499	71.499	71.427
	9. The Appliances Energy data set				10. SuperConductivity			
Methods	PCA	SPCA	RSPCA	ROBSPCA	PCA	SPCA	RSPCA	ROBSPCA
The number of factors/ PC_s	3	3	3	3	2	2	2	2
Cumul. Eigen. percentage (%)	79.3	78.7	78.7	78.8	92.3	91.9	91.9	90.5
Maximum lag of variables	6	6	6	6	10	10	10	10
RMSE	101.74	101.76	101.76	101.75	27.314	27.332	27.332	27.319

Extracted from Table 2, Table 3 compares the dimensionality reduction performance of the PCA, SPCA, RSPCA, and ROBCPCA methods performed under the ideal conditions as presented above. To show the dimensionality reduction performance for all ten experimental datasets in the same Figure, the article stretched the data across these 10 data sets by multiplying their data in Table 3 with coefficients 1000, 1, 1, 15, 10, 10, 20, 20 and 100, respectively. Figure 1 contains the column histograms of the stretched data sets.

Table 3. The variable dimension reduction performance of the methods (RMSE)

Methods	D1	D2	D3	D4	D5	D6	D7	D8	D9	D20
PCA	**0.190**	1.484	**715.961**	1152.395	161.415	**91.824**	365.970	71.459	**101.742**	**27.314**
SPCA	0.197	**1.066**	826.276	1152.531	**161.441**	309.241	85.467	71.499	101.764	27.332
RSPCA	0.197	1.067	1373.567	1152.531	**161.441**	309.241	85.467	71.499	101.764	27.332
ROBSPCA	0.205	**1.066**	2642.834	**1151.247**	161.441	309.235	**85.462**	**71.427**	101.747	27.319

In this table, the codes from D1 to D10 are assigned, respectively, to the ten experimental data sets in Table 1 from top to bottom.

Table 3 and Fig. 1 clearly show that in the ten experimental data sets, there are 5/10 cases, where the dimensionality reduction performance of the PCA method is higher than that of the family of the SPCA method, and the remaining 5/10 cases are the opposite, that is, the dimensionality reduction performance of the family of the SPCA method is higher than that of the PCA methods. However, there are many cases where the difference of the dimensionality reduction performance is very insignificant. Namely, there are 5/10 cases (corresponding to Residence Buiding, S&P 500, Air Quality, Appliances Energy, and SuperConductivity datasets) where the dimensionality reduction performance of the PCA, SPCA, RSPCA, and ROBSPCA methods are considered the same.

Table 3 and Fig. 1 also show that in the remaining 5/10 cases, there are 3/10 cases where the dimensionality reduction performance of the PCA method is much higher than that of the family of SPCA methods (corresponding to the VN30, CPI, and NASDAQ datasets). The 2/10 remaining cases where the dimensionality reduction performance of the family of SPCA methods is much higher than that of the PCA method (corresponding to the VIP and DJI data sets).

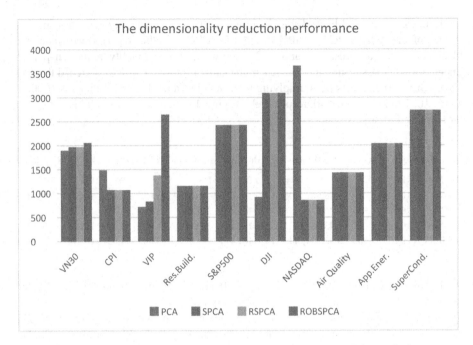

Fig. 1. Compare the dimensionality reduction performance of the methods

In other words, the experiments show that the dimensionality reduction performance of the PCA method and the family of the SPCA method is competitive.

5 Conclusion

This article has experimentally compared the dimensionality reduction performance of the SPCA methods family with the PCA method on ten real-world data sets under ideal conditions. The experimental results show that the dimensionality reduction performance of the SPCA methods family is not higher than the dimensionality reduction performance of the PCA method as long thought by researchers. Their dimensionality reduction performance is competitive.

However, the family of SPCA methods is still worthy of application interest because each principal component found by these methods is only a linear combination of several original predictors. That allows evaluating the impact of some original predictors on the target variable in the context of too many original predictors that can potentially affect the change of the target variable.

Funding. This research did not receive any specific grant from funding agencies in the public, commercial, or not-for-profit sectors.

References

1. Kapetanios, G., Papailias, F., et al.: Big data and macroeconomic nowcasting: methodological review. Economic Statistics Centre of Excellence, National Institute of Economic and Social Research (2018)
2. Chandrashekar, G., Sahin, F.: A survey on feature selection methods. Comput. Electr. Eng. **40**(1), 16–28 (2014). https://doi.org/10.1016/j.compeleceng.2013.11.024
3. Do Van, T.: Modeling stock price forecasting in the context of high dimensional data. In: Proceedings of the 10th National Science and Technology Conference, FAIR, Da Nang, 17–18 August 2017, pp. 422–433 (2017). (in Vietnamese). https://doi.org/10.15625/vap.2017.00051
4. Do Van, T., Hai, N.M., Hieu, D.D.: Building an unconditional forecast model of Stock Market Indexes using combined leading indicators and principal components: application to the Vietnamese Stock Market. Indian J. Sci. Technol. **11**(2), 1–13. https://doi.org/10.17485/ijst/2018/v11i2/104908
5. Shlens, J.: A tutorial on principal component analysis. arXiv preprint arXiv:1404.1100 (2014)
6. Kim, H.H., Swanson, N.R.: Mining big data using a parsimonious factor, machine learning, variable selection, and shrinkage methods. Int. J. Forecast. **34**(2), 339–354 (2018)
7. Zou, H., Hastie, T., Tibshirani, R.: Sparse principal component analysis. J. Comput. Graph. Stat. **15**(2), 265–286 (2006)
8. Chikamatsu, K., Hirakata, N., Kido, Y., Otaka, K., et al.: Nowcasting Japanese GDPs. Bank of Japan (2018)
9. Maaten, L.V.D., Postma, E.: Dimensionality reduction: a comparative review. J. Mach. Learn. Res. **10**, 66–71 (2009)
10. Sorzano, C.O.S., Vargas, J., Montano, A.P.: A survey of dimensionality reduction techniques, pp. 1–35. arXiv preprint arXiv:1403.2877 (2014)
11. Sarveniazi, A.: An actual survey of dimensionality reduction. Am. J. Comput. Math. **04**(02), 55–72 (2014). https://doi.org/10.4236/ajcm.2014.42006
12. Erichson, N.B., Zheng, P., Manohar, K., Brunton, S.L., Kutz, J.N., Aravkin, A.Y.: Sparse principal component analysis via variable projection. SIAM J. Appl. Math. **80**(2), 977–1002 (2020)
13. Koop, G., Quinlivan, R.: Analysis of Economic Data, vol. 2. Wiley, Chichester (2020)
14. Ghysels, E., Santa-Clara, P., Valkanov, R.: The MIDAS touch: mixed data sampling regression models, 34 p. Powered by the California Digital Library University of California (2004). https://escholarship.org/uc/item/9mf223rs
15. Ghysels, E., Santa-Clara, P., Valkanov, R.: Predicting volatility: getting the most out of return data sampled at different frequencies. J. Econom. **131**(1–2), 59–95 (2006)
16. Greene, W.H.: Econometric Analysis, 7th edn. New York University/Prentice-Hall (2012)
17. UCI-Machine Learning Repository. archive.ics.uci.edu/ml/datasets.php. Accessed 5 May 2021
18. Rafiei, M.H., Adeli, H.: A novel machine learning model for estimation of sale prices of real estate units. UCI Machine Learning Repository (2016)
19. Hoseinzade, E., Haratizadeh, S.: CNNpred: CNN-based stock market prediction using a diverse set of variables. UCI Machine Learning Repository (2019)
20. De Vito, S., Massera, E., Piga, M., Martinotto, L., Di Francia, G.: Air quality. UCI Machine Learning Repository (2016)
21. Candanedo, L.M., Feldheim, V., Deramaix, D.: Appliances energy prediction data set. UCI Machine Learning Repository (2017)

22. Hamidieh, K.: Superconductivity data. UCI Machine Learning Repository (2018)
23. Zhang, Y., Li, S., Teng, Y.: Dynamic processes monitoring using recursive kernel principal component analysis. Chem. Eng. Sci. **72**, 78–86 (2012)
24. Erichson, N.B., Zheng, P., Aravkin, S.: Sparsepca: Sparse Principal Component Analysis (SPCA). R package version 0.1.2 (2018). https://CRAN.R-project.org/package=sparsepca
25. Karatzoglou, A., Smola, A., Hornik, K., Zeileis, A.: Kernlab-an S4 package for kernel methods in R. J. Stat. Softw. **11**(9), 1–20 (2004). URL http://www.jstatsoft.org/v11/i09/
26. Kuhn, M., et al.: Building predictive models in R using the caret package. J. Stat. Softw. **28** (5), 1–26 (2008)

A Game Theory Approach for Water Exchange in Eco-Industrial Parks: Part 1 - A Case Study Without Regeneration Units

Kien Cao Van[✉]

Faculty of Information Technology, Nguyen Tat Thanh University,
Ho Chi Minh City, Vietnam
cvkien@ntt.edu.vn

Abstract. Eco-Industrial Parks (EIP) offer to enterprises the possibility to make economic benefits and to minimize environmental impacts by sharing flows and increasing inter-enterprise exchanges. Part 1 of the paper presents a mathematical programming formulation for the optimal design of water exchange networks in an EIP without regeneration units, based on a single-leader multi-follower (SLMF) methodology. The goal of each follower is to minimize his operating cost, while the leader's goal is to reduce the consumption of freshwater in the EIP. Examples of EIPs are studied numerically. The results show that the SLMF methodology is very reliable in the multi-criteria scenarios of the EIP design, providing numerical Nash equilibrium solutions.

Keywords: Game theory · Nash equilibrium · Single-leader multi-follower game · Mixed-integer programming · Eco-Industrial Park

1 Introduction

In recent years, the term "climate change" has received a lot of attention. It not only adversely affects the lives of people at the present, but also threatens the environment in the future. The main cause of the Earth's climate change is the increase in activities that generate greenhouse gas waste, excessive exploitation of natural resources in industry. To protect the global environment as well as increase economic benefits, the concept of *Industrial Ecology* (IE) has emerged [2]. IE provides an innovative way to produce goods and services based on a circular model. Indeed, in these industrial ecosystems, resource consumption and waste generation are minimized by allowing the waste materials from one industry to serve as raw material for another. An application to this concept are EIP [7]: "A community of manufacturing and service businesses seeking enhanced environmental and economic performance through collaboration in managing environmental and resource issues including energy, water, and materials. By working together, the community of businesses seeks a collective benefit that

Published by Springer Nature Switzerland AG 2021. All Rights Reserved
P. Cong Vinh and N. Huu Nhan (Eds.): ICTCC 2021, LNICST 408, pp. 149–161, 2021.
https://doi.org/10.1007/978-3-030-92942-8_13

is greater than the sum of the individual benefits each company would realize if it optimized its individual performance only". To achieve these objectives, optimization methods are required to design optimal exchange networks between enterprises. However, a major challenge in developing optimization methods is to deal with the network complexity of the solutions, especially when a lot of interconnections are considered, as is the case with EIP.

Modeling EIPs is somewhat a complex problem due to its size and the number of objectives to take into account. In the literature, there are two main approaches for designing and optimizing water-exchange networks in the EIP: multi-objective optimization (MOO) on one hand and game theory on the other hand. The MOO approach requires enterprises to coordinate their strategies and share information. This requirements usually does not exist in the optimal design of the exchange networks in the EIP as it indicated in the works of [9] and [11]. We refer the reader to [2–4,8] for the survey on MOO approach.

Another approach to design an optimal resource exchange network in an eco-industrial park is game theory. More precisely, the concept of the single-leader multi-follower game problem. Indeed, at the upper-level problem, there exists an EIP authority that wants to minimize resource consumption and generation of waste, while at the lower-level problem, each enterprise tries to minimize his operating cost. In SLMF game, the EIP authority makes decision first by anticipating the response of the enterprises. At the same time, all enterprises compete with each other in a generalized Nash game with the strategies of the EIP authority as exogenous parameters. It's worth mentioning that, at the lower level, enterprises play a generalized Nash equilibrium between them, so enterprises involved would be able to protect confidential information, without the need to share information with the other enterprises of the park. The existence of an optimal solution satisfying Nash equilibrium would ensure that none of each enterprise is prejudiced compared to others. We refer the reader to [6] for a survey of generalized Nash equilibrium problems and to [1,9–11] for a survey of SLMF approach.

In this work, the design and optimization of industrial water networks in eco-industrial parks are studied by formulating and solving single-leader multi-follower game problems. The approach is validated on a case study of water integration in EIP without regeneration units. The result shown that the equilibrium solution will provide economic benefits to enterprises participating in the EIP.

The remainder of this paper is organized as follows: Sect. 2 is dedicated the methodology and model formulation which briefly describes the problem addressed in this article and present a model for water exchange networks in EIPs without regeneration units, based on a single-leader-multiple-follower model. The reformatting of the EIP modeling problem as a mixed integer linear programming problem is addressed in Sect. 3. Section 4 is dedicated to numerical experiments on reasonably large EIP. Finally, general conclusions and future perspectives are presented in Sect. 5.

2   Methodology and Model Formulation

2.1   Problem Statement

Let n denote the given number of enterprises, $I := \{1, \ldots, n\}$ denote the index set of enterprises, and 0 denote the sink node. The sink node is a place to store contaminated wastewater. Thus, we define $I_0 = \{0\} \cup I$. Each enterprise has its own pre-defined water input requirement and quality characteristics, as well as the quantity and quality of available output wastewater. For each enterprise, the resource consumption can be freshwater, and/or wastewater from other enterprises. Indeed, the polluted water from an enterprise can be sent to the sink node, and/or sent to other enterprises. The objective of the model is to determine a network of connections of water streams among them so that both the total freshwater consumption and the annualized operating cost of each enterprise in the park are minimized, while satisfying all process and environmental constraints.

2.2   Minimizing Operating Costs

Each enterprise $i \in I$ receives the wastewater from other enterprises in the EIP. Nevertheless, for technical constraints on the process P_i, the pollutant concentration delivered by the other enterprises cannot exceed a certain maximum value denoted here by $C_{i,\text{in}}$ [ppm]. On the other hand enterprise i has a given contaminant load M_i [g/h] that needs to be diluted before exiting the enterprise. To do so, enterprise i needs to use an amount of fresh water z_i [T/h] such that the outlet pollutant concentration is less than a limit concentration $C_{i,\text{out}}$ [ppm]. Actually considering that enterprise i will optimize his process, it is assumed that each enterprise $i \in I$ will only consumes the exact amount of freshwater it needs to satisfy $C_{i,\text{out}}$, and therefore, its output pollutant concentration is always equal to this constant.

There is an exchange of materials between the enterprises in EIPs. On the other hand, let E be the configuration for the water exchange network in EIPs such that $(i,j) \in E$ then the enterprise i can send his wastewater to enterprise j. Especially if the enterprise i uses the connection $(i,0)$, it means that it is discharging polluted water outside the park.

Defining the *stand-alone* and *complete* configuration, respectively, as follows

$$E_{\text{st}} := \{(i,0) : i \in I\} \quad \text{and} \quad E_{\max} := \{(i,j) : i \in I, j \in I_0\},$$

thus a valid configuration E must satisfy that $E_{\text{st}} \subset E \subset E_{\max}$. We denote by $\mathcal{E}$ the set of all valid configurations for the EIP. Furthermore, for any $E \in \mathcal{E}$, we denote by $E^c = E_{\max} \setminus E$ the family of connections that are not present in E.

In term of variables, each enterprise $i \in I$ sends polluted water to $j \in I_0$, taken into account by variable $F_{i,j}$ [T/h]. In addition, we set $F = (F_{i,j} : (i,j) \in E_{\max})$, the full vector of fluxes through the configuration.

Furthermore, for each enterprise $i \in I$, we denote $F = (F_i, F_{-i})$ where $F_i = (F_{i,j} : j \in I)$ and $F_{-i} = (F_{k,j} : k \in I \setminus \{i\}, j \in I)$, to emphasize the vector of fluxes between enterprise i. Then, for a fixed network E, the EIP model without regeneration units must satisfy the following constraints:

1. Water mass balance constraint for an enterprise $i \in I$:

$$z_i + \sum_{(k,i)\in E} F_{k,i} = \sum_{(i,j)\in E, j\neq 0} F_{i,j} + F_{i,0}. \tag{1}$$

2. Contaminant mass balance constraint for an enterprise $i \in I$:

$$M_i + \sum_{(k,i)\in E} C_{k,\text{out}} F_{k,i} = C_{i,\text{out}} \left(\sum_{(i,j)\in E, j\neq 0} F_{i,j} + F_{i,0} \right). \tag{2}$$

3. Inlet/outlet concentration constraints for an enterprise $i \in I$:

$$\sum_{(k,i)\in E} C_{k,\text{out}} F_{k,i} \leq C_{i,\text{in}} \left(z_i + \sum_{(k,i)\in E} F_{k,i} \right). \tag{3}$$

4. Positivity of fluxes and null fluxes outside the connections: we need all the fluxes to be positive:

$$\begin{cases} F_{i,j} \geq 0, & \forall (i,j) \in E \\ z_i \geq 0, & \forall i \in I. \end{cases} \tag{4}$$

Of course, we also put

$$\forall (i,j) \in E^c, \ F_{i,j} = 0, \tag{5}$$

that is, the effective fluxes can only pass through the existing connections in E.

By combining Eqs. (1) and (2) we obtain:

$$M_i + \sum_{(k,i)\in E} C_{k,\text{out}} F_{k,i} = C_{i,\text{out}} \left(z_i + \sum_{(k,i)\in E} F_{k,i} \right), \quad \forall i \in I. \tag{6}$$

From the Eq. (6) the freshwater consumption of the enterprise $i \in I$ is defined by

$$z_i(F_{-i}) = \frac{1}{C_{i,\text{out}}} \left(M_i + \sum_{(k,i)\in E} (C_{k,\text{out}} - C_{i,\text{out}}) F_{k,i} \right). \tag{7}$$

Thus, each enterprise i wants to minimize his operating cost, given by

$$\text{Cost}_i(F_i, F_{-i}, E) = A \left[c \cdot z_i(F_{-i}) + \gamma_{i,0} F_{i,0} + \sum_{(k,i)\in E} \gamma_{k,i} F_{k,i} + \sum_{(i,j)\in E, j\neq 0} \gamma_{i,j} F_{i,j} \right], \tag{8}$$

where A [h] stands for the annual EIP operating hours, c [$/T] for the purchase price of freshwater, $\gamma_{i,0}$[$/T] for the price of wastewater discharge, and $\gamma_{p,q}$[$/T] for the cost of sending wastewater from enterprise p to q.

With all these considerations, each enterprise's $i \in I$ optimization problem is given by $P_i\,(F_{-i}, E)$

$$\min_{F_i}\ \mathrm{Cost}_i\,(F_i, F_{-i}, E)$$

$$s.t. \begin{cases} \text{Equations (1)-(3)-(5),} \\ z_i(F_{-i}) \geq 0, \\ F_i \geq 0. \end{cases} \tag{9}$$

For a network $E \in \mathcal{E}$, the family of equilibria for E at the lower-level problem is given by $\mathrm{Eq}(E)$. Furthermore,

$$F \in \mathrm{Eq}(E) \iff \forall i \in I,\ F_i \text{ solves the problem } P_i\,(F_{-i}, E)\,.$$

Remark 1. *If no enterprise sends wastewater to enterprise $i \in I$, the amount of freshwater consumed by enterprise i must be*

$$z_i = \frac{M_i}{C_{i,\mathrm{out}}}.$$

Then, the cost of stand-alone configuration is given by

$$\mathrm{STC}_i = A \cdot (c + \gamma_{i,0})\frac{M_i}{C_{i,\mathrm{out}}}.$$

2.3   Minimizing Consumption of Natural Resources

In the model, the EIP authority tries to optimize the total freshwater consumption, and so he wants to minimize the objective function

$$Z(F) = \sum_{i \in I} z_i(F_{-i}). \tag{10}$$

The authority must ensure a relative improvement of $\alpha \in\,]0,1[$ in the costs, with respect to the stand-alone operation, that is,

$$\mathrm{Cost}_i(F_i, F_{-i}, E) \leq \alpha \cdot \mathrm{STC}_i \quad \forall i \in I. \tag{11}$$

Thus, the optimization problem of the EIP authority is given by

$$\min_{F \in \mathbb{R}^{|E_{\max}|}, E \in \mathcal{E}}\ Z(F)$$

$$s.t. \begin{cases} F \in \mathrm{Eq}(E), \\ \mathrm{Cost}_i(F_i, F_{-i}, E) \leq \alpha \cdot \mathrm{STC}_i, \quad \forall i \in I. \end{cases} \tag{12}$$

3 Mixed-Integer Programming Reduction

As we can observe the EIP authority problem (12) is mathematical programming with equilibrium constraints, which is hard to solve. In the following, however, this type of problem will be reformulated as a single mixed-integer programming problem.

3.1 Characterization of Equilibria

In order to characterize the equilibrium set $Eq(E)$ as a system of equalities and inequalities, for each enterprise $i \in I$, we define the set

$$E_{i,\text{act}} := \left\{ (i,j) \in E \ : \ \gamma_{i,j} = \gamma_i^* := \min_{(i,k) \in E} \gamma_{i,k} \right\}. \tag{13}$$

Theorem 2. *For any valid exchange network $E \in \mathcal{E}$ and denoting $S(E)$ by the set*

$$S(E) = \left\{ F \ : \ \forall i \in I, \begin{array}{l} z_i + \displaystyle\sum_{(k,i) \in E} F_{k,i} = \sum_{(i,j) \in E} F_{i,j} \\[2mm] \displaystyle\sum_{(k,i) \in E} C_{k,\text{out}} F_{k,i} \le C_{i,\text{in}} \left(z_i + \sum_{(k,i) \in E} F_{k,i} \right) \\[2mm] z_i(F_{-i}) \ge 0 \\ F_i \ge 0 \\ F_i \big|_{E_{i,\text{act}}^c} = 0 \end{array} \right\} \tag{14}$$

then, one has $S(E) = Eq(E)$. Furthermore, any optimal solution (F, E) of the mathematical programming problem

$$\min_{F \in \mathbb{R}^{|E_{\max}|}, E \in \mathcal{E}} Z(F)$$

$$s.t. \begin{cases} F \in S(E), \\ \text{Cost}_i(F_i, F_{-i}, E) \le \alpha \cdot \text{STC}_i, & \forall i \in I. \end{cases} \tag{15}$$

is an optimal solution of the SLMF problem (12).

For the proof of Theorem 2, we prefer the reader to Appendix A in Part 2 of the paper [5].

3.2 Mixed-Integer Formulation

Constraint $F_i \big|_{E_{i,\text{act}}^c} = 0, \forall i \in I$ depends on the configuration of the water exchange network and is therefore difficult to implement numerical experiments. Thus, we will reduce the single optimization problem (12) to the mixed-integer programming problem.

Let $(i,j) \in E_{\max}$, we define

$$C(i,j) := \{(i,k) \in E_{\max} : \gamma_{i,k} = \gamma_{i,j}\} \tag{16}$$

the *arc class* of (i,j). Furthermore, we denote by $\mathcal{C}_i = \{C(i,j) : (i,j) \in E_{\max}\}$ the family of all arc classes of enterprise i.

If there exists one class $C(i,j) \in \mathcal{C}_i$ such that $E_{i,\text{act}} \subseteq C(i,j)$, then we will call it the *active class* of E of the enterprise i, and we will denote it by $C_i(E)$.

Now, let $D = \bigcup_{i \in I} \mathcal{C}_i$, the set of all arc classes of enterprises. We introduce the boolean variable $y = (y_C)_{C \in D} \in \{0,1\}^{|D|}$ in the following way: for each enterprise $i \in I$ and each arc class $C \in \mathcal{C}_i$, we set

$$y_C = \begin{cases} 1 & \text{if } C \text{ is the active class of } i, \\ 0 & \text{otherwise.} \end{cases}$$

With this new boolean variable, we set the following constraints:

1. For each enterprise $I \in I$,

$$\sum_{C \in \mathcal{C}_i} y_C = 1, \tag{17}$$

namely, only one class is active.

2. For each enterprise $I \in I$,

$$\sum_{(i,j) \in C} F_{i,j} \leq K \cdot y_C, \quad \forall C \in D, \tag{18}$$

where K is a constant large enough. This constraint ensures that, if $(i,j) \notin C$, then $F_{i,j} = 0$.

From the new boolean variable $y \in \{0,1\}^{|D|}$, we set up the graph associated to y as

$$E(y) = \left(\bigcup\{C : y_C = 1\}\right) \cup \{(i,0) : i \in I\}. \tag{19}$$

Now, let's consider the following Mixed-Integer optimization problem:

$$\min_{F \in \mathbb{R}^{|E_{\max}|}, y \in \{0,1\}^{|D|}} Z(F) \tag{20}$$

$$s.t. \begin{cases} \text{Equations (1)-(3)-(4)-(11)-(17)-(18).} \end{cases}$$

Theorem 3. *If (F, E) is an optimal solution of problem (15), then (F, y^E) is an optimal solution of problem (20), where $y^E \in \{0,1\}^{|D|}$ is defined as*

$$y_C^E = \begin{cases} 1 & \text{if } C = C_i(E) \text{ for some } i \in I, \\ 0 & \text{otherwise.} \end{cases}$$

If (F, y) is an optimal solution of problem (20), then $(F, E(y))$ is an optimal solution of problem (15).

For the proof of Theorem 3, we prefer the reader to Appendix B in Part 2 of the paper [5].

3.3   Null Class as Exit Option

We can observe that the stand-alone configuration E_{st} is always a feasible configuration for the EIP model. However, with the constraint (11), the problem may become infeasible. Therefore, we need to take into account the possibility of excluding some enterprises from the network when the EIP authority does not ensure to satisfy constraint (11) for such enterprises.

Now, for each enterprise $i \in I$, we introduce a boolean variable $y_{i,\text{null}} \in \{0,1\}$ such that

$$y_{i,\text{null}} = \begin{cases} 1 & \text{if } i \text{ breaks the constraint (11),} \\ 0 & \text{otherwise.} \end{cases}$$

With this boolean variable, we will add the following constraints to problem (20).

1. For each enterprise $i \in I$,

$$y_{i,\text{null}} + \sum_{C \in \mathcal{C}_i} y_C = 1, \qquad (21)$$

2. For each enterprise $i \in I$,

$$\sum_{(i,j) \in C(i,0)} F_{i,j} \leq K \cdot (y_{C(i,0)} + y_{i,\text{null}}), \qquad (22)$$

$$\sum_{(i,j) \in E_{\max}, j \neq 0} F_{i,j} \leq K \cdot (1 - y_{i,\text{null}}). \qquad (23)$$

Constraints (22) and (23) are to ensure that, if the enterprise breaks the constraint, then the enterprise e do not share polluted water with other enterprises and will use the connection $(i,0)$.

3. For each enterprise $i \in I$,

$$\sum_{(k,i) \in E_{\max}} F_{k,i} \leq K \cdot (1 - y_{i,\text{null}}). \qquad (24)$$

This constraint is to ensure that, if the enterprise breaks the constraint (11), then no enterprises can send him any polluted water.

4. For each enterprise $i \in I$,

$$\text{Cost}_i(F_i, F_{-i}, E(y)) \leq \alpha_i \cdot \text{STC}_i \cdot (1 - y_{i,\text{null}}) + \text{STC}_i \cdot y_{i,\text{null}}. \qquad (25)$$

We denote by $\overline{D} = D \cup \{\text{Null}_i : i \in I\}$, where Null_i is the null class, associated to $y_{i,\text{null}}$, and $D_0 = D \setminus \{C(i,0) : i \in I\}$. Denoting

$$\text{STC}_i(y_{i,\text{null}}) := \alpha_i \cdot \text{STC}_i \cdot (1 - y_{i,\text{null}}) + \text{STC}_i \cdot y_{i,\text{null}}.$$

With all the foregoing, problem (20) becomes

$$\min_{F \in \mathbb{R}^N, y \in \{0,1\}^{|\overline{D}|}} Z(F)$$

$$s.t. \begin{cases} \text{Equations (1)-(3)-(4)-(11)-(21)-(22)-(23)-(24),} & (26) \\ \sum_{(i,j) \in C} F_{i,j} \leq K \cdot y_C, & \forall C \in D_0. \end{cases}$$

4   Numerical Experiments

4.1   Case Study

Now, we simulate numerical examples of the model described in Sect. 2. We assume that

$$\gamma_{i,j} = \begin{cases} \delta & \text{if } j \in I, \\ \beta & \text{if } j = 0. \end{cases} \tag{27}$$

From (27), we can observe that each enterprise i has to pay for both water receipt and water deposit. Therefore, for each enterprise $i \in I$, the set $\mathcal{C}_i = \{C_{i,p}, C_{i,0}\}$ where

$$C_{i,p} = \{(i,j) \in E_{\max} \: : \: j \in I\} \quad \text{and} \quad C_{i,0} = \{(i,0)\}.$$

Now, for each enterprise $i \in I$, we introduce three integer variables, $y_{i,p}, y_{i,0},$ $y_{i,\text{null}} \in \{0,1\}$ as follows:

- If $y_{i,p} = 1$, it means the connections in $C_{i,p}$ are included in the network.
- If $y_{i,0} = 1$, it means the connection $(i,0)$ is the only exit connection for i, and i participates in the EIP.
- If $y_{i,\text{null}} = 1$, it means the connection $(i,0)$ is the only exit connection for i, and i does not participate in the EIP.

Note that only one of these integer variables takes the value 1, i.e., $y_{i,\text{null}} + y_{i,p} + y_{i,0} = 1, \forall i \in I$, and in doing so it determines the network E to be implemented and the operation that each enterprise can do within this network. We denote by $y \in \{0,1\}^{3n}$ the vector of all integer variables of all enterprises.

Since the optimization problem (26) can have several solutions and to get a solution with more enterprises involved, we replace $Z(F)$ by

$$Z(F) + \text{Coef} \cdot \sum_{i \in I} y_{i,\text{null}}, \tag{28}$$

where Coef ≥ 0 is a coefficient to penalize the objective function. Then, the optimization problem (26) becomes:

$$\min_{F,y} \quad Z(F) + \text{Coef} \cdot \sum_{i \in I} y_{i,\text{null}}$$

$$\text{s.t.} \begin{cases} \text{Equations (1)-(3)-(4)-(11)}, & \\ y_{i,\text{null}} + y_{i,p} + y_{i,0} = 1, & \forall i \in I, \\ \displaystyle\sum_{(i,j) \in C_{i,p}} F_{i,j} \leq K \cdot y_{i,p}, & \forall i \in I, \\ F_{i,0} \leq K \cdot (y_{i,0} + y_{i,\text{null}}), & \forall i \in I, \\ \displaystyle\sum_{(i,j) \in E_{\max}, j \neq 0} F_{i,j} \leq K \cdot (1 - y_{i,\text{null}}), & \forall i \in I, \\ \displaystyle\sum_{(k,i) \in E_{\max}} F_{k,i} \leq K \cdot (1 - y_{i,\text{null}}), & \forall i \in I. \end{cases} \tag{29}$$

We use input data on the scale of an EIP of 15 enterprises. The data is given in Table 1 and Table 2. Data is partially inspired from [4]. Additionally, it is supposed that the EIP operates $A = 1\,$h.

Table 1. Parameters of the network.

Enterprise i	$C_{i,\text{in}}$ (ppm)	$C_{i,\text{out}}$ (ppm)	M_i (g/h)
1	0	50	1000
2	0	100	2500
3	10	50	1500
4	10	100	5000
5	30	300	20000
6	40	600	5000
7	20	200	5000
8	50	400	15000
9	50	100	10000
10	50	600	20000
11	110	450	15000
12	200	400	10000
13	500	1100	30000
14	300	3500	15000
15	600	2500	25000

Table 2. Associated costs.

Parameter	Value ($/tonne)
c	0.13
β	0.22
δ	0.01

All simulations below performed using Julia `Julia v1.0.5` programming language, using `CPLEX V12.10.0` as a solver.

4.2   Results and Discussion

In this section, we present the numerical results with the above data. The resulting optimized EIP network is shown in Fig. 1, and it corresponds to $\alpha = 0.90$ and Coef $= 1$. This optimal network provides operating cost of each enterprise and total freshwater consumption that are lower than a stand-alone network as shown in Table 3.

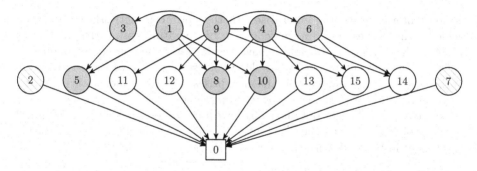

Fig. 1. The optimal configuration in the case $\alpha_i = 0.90$ and Coef $= 1$. Dashed nodes are operating in stand-alone node and gray nodes are active consuming freshwater.

Table 3. Summary of results of the EIP.

Enterprise	Freshwater stand-alone (T/h)	Freshwater in EIP (T/h)	Cost$_i$ stand-alone (MMUSD/hour)	Cost$_i$ in EIP (MMUSD/hour)	% Reduction in Cost$_i$
1	20.00	20.00	7.00	2.80	60.00
2	25.00	25.00	8.75	8.75	0.00
3	30.00	33.75	10.50	4.80	54.28
4	50.00	50.00	17.50	7.11	59.36
5	66.67	35.13	23.33	21.00	10.00
6	8.33	5.36	2.92	0.82	71.84
7	25.00	25.00	8.75	8.75	0.00
8	37.50	16.74	13.12	11.81	10.00
9	100.00	100.00	35.00	14.00	60.00
10	33.33	13.74	11.67	10.01	14.18
11	33.33	0.00	11.67	9.86	15.51
12	25.00	0.00	8.75	7.67	12.38
13	27.27	0.00	9.54	6.90	27.71
14	4.28	0.00	1.50	1.08	28.12
15	10.00	0.00	3.50	2.73	21.89
Total	495.72	324.72	173.50	118.09	31.94

For this case study, when the enterprises operate stand alone, then the entire system consumes a total 495.72 (T/h) of freshwater. The optimal design obtained with SLMF approach allowed to reduce its freshwater requirement to 324.72 (T/h) which is equivalent to a reduction of 34.5%. Furthermore, the water demand of enterprises 11, 12, 13, 14, and 15 are entirely supplied by other enterprises.

When working in the EIP, the EIP authority ensures that each enterprise gets at least a 10% reduction in costs. Total operating cost is reduced compared to the stand-alone case, as expected, from 173.50 ($/h) to 118.09 ($/h), i.e., 31.94% reduction. Nevertheless, exploring Table 3, we can observe that the cost reduction is not uniform across the enterprises. In the optimized network, enterprise 6 achieves the highest percentage reduction of operating cost corresponding to 71.84% while enterprises 5 and 8 have the lowest reduction corresponding to 10.00% with respect to the stand-alone configuration. Nevertheless, the EIP authority cannot guarantee that the enterprises 2 and 7 reduce their operating costs by at least 10%, so they operate stand-alone.

Moreover, the benefit obtained varies with the different values of α. Figure 2 is a box plot to describe the distribution of the cost reduction of participating enterprises in the park with parameters α in the interval $[0.80, 0.99]$.

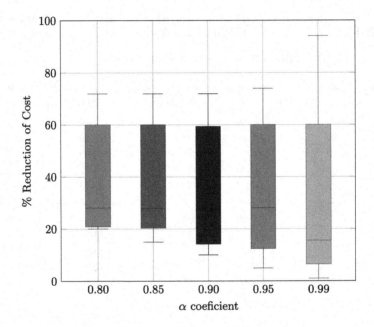

Fig. 2. % Reduction of cost with $\alpha \in [0.80, 0.99]$ and Coef = 1.

A box plot is a standardized way of displaying the distribution of data based on a five number summary: lower limit, lower quartile (Q_1: 25th Percentile), median (Q_2: 50th Percentile), upper quartile (Q_3: 75th Percentile), and upper limit. Let us choose $\alpha = 0.90$ to analyze the distribution of the cost reduction of enterprises, and other parameters α can be analyzed analogously. All enterprises will reduce less than or equal to 71.84% (upper limit). All enterprises can reduce operating costs by at least 10% (lower limit). At least 75% of enterprises can reduce costs by 14.18% or higher. The lower (resp. upper) quartile Q_1 (resp. Q_3) is 14.18% (resp. 59.36%) which is equivalent to that there are 25% (resp. 75%) of enterprises can reduced their operating costs less than or equal to 14.18% (resp. 59.36%), while the median reduction Q_2 in this case is 27.71%, so exactly haft the enterprises are reduced lesser or higher 27.71% compared to their stand-alone operating costs.

5   Conclusion and Perspectives

In this work, we have presented a game theory methodology to optimize the water exchange networks in the EIP. The approach is validated on a case study of water integration in the EIP without regeneration units. Using this methodology the results show that the game theory methodology is very reliable in the multi-criteria scenarios of the EIP design. More precisely, the EIP authority ensures that each enterprise gets at least a 10% reduction in costs on one hand, on the other hand the total freshwater consumption and total operating cost in the optimal configuration have been reduced by 34.5% and 31.94%, respectively.

A methodology taking into account in this present work only the single contaminant case is implemented. Thus, we would like to address the EIP design with multi contaminant case in the future.

Acknowledgments. This research is funded by Nguyen Tat Thanh University, Ho Chi Minh city, Vietnam.

References

1. Aviso, K.B., Tan, R.R., Culaba, A.B., Cruz, J.B.: Bi-level fuzzy optimization approach for water exchange in eco-industrial parks. Process Saf. Environ. Prot. **88**(1), 31–40 (2010)
2. Boix, M., Montastruc, L., Azzaro-Pantel, C., Domenech, S.: Optimization methods applied to the design of eco-industrial parks: a literature review. J. Clean. Prod. **87**, 303–317 (2015)
3. Boix, M., Montastruc, L., Pibouleau, L., Azzaro-Pantel, C., Domenech, S.: Eco industrial parks for water and heat management. Comput. Aided Chem. Eng. **29**, 1175–1179 (2011)
4. Boix, M., Montastruc, L., Pibouleau, L., Azzaro-Pantel, C., Domenech, S.: A multiobjective optimization framework for multicontaminant industrial water network design. J. Environ. Manage. **92**(7), 1802–1808 (2011)
5. Cao Van, K.: A game theory approach for water exchange in eco-industrial parks: Part 2 - a case study with regeneration units. Accepted to 7th EAI International Conference on Nature of Computation and Communication, p. 17pp (2021)
6. Facchinei, F., Kanzow, C.: Generalized Nash equilibrium problems. Ann. Oper. Res. **175**, 177–211 (2010)
7. Lowe, E.A.: Creating by-product resource exchanges: strategies for eco-industrial parks. J. Clean. Prod. **5**(1), 57–65 (1997)
8. Montastruc, L., Boix, M., Pibouleau, L., Azzaro-Pantel, C., Domenech, S.: On the flexibility of an eco-industrial park (EIP) for managing industrial water. J. Clean. Prod. **43**, 1–11 (2013)
9. Ramos, M., Boix, M., Aussel, D., Montastruc, L., Domenech, S.: Water integration in eco-industrial parks using a multi-leader-follower approach. Comput. Chem. Eng. **87**, 190–207 (2016)
10. Ramos, M., Rocafull, M., Boix, M., Aussel, D., Montastruc, L., Domenech, S.: Utility network optimization in eco-industrial parks by a multi-leader-follower game methodology. Comput. Chem. Eng. **112**, 132–153 (2018)
11. Salas, D., Van Cao, K., Aussel, D., Montastruc, L.: Optimal design of exchange networks with blind inputs and its application to eco-industrial parks. Comput. Chem. Eng. **143**, 107053 (2020)

A Game Theory Approach for Water Exchange in Eco-Industrial Parks: Part 2 - A Case Study with Regeneration Units

Kien Cao Van[✉]

Faculty of Information Technology, Nguyen Tat Thanh University,
Ho Chi Minh City, Vietnam
cvkien@ntt.edu.vn

Abstract. Part 2 of the paper presents the optimal design of the water exchange network in the eco-industrial parks (EIP) using the single-leader multi-follower (SLMF) game methodology. The SLMF game methodology is suitable on the case study of water management in EIP with regeneration units. The main goal of regeneration units is to reduce the contaminant concentration with different quality specifications. From there, the enterprises can not only use the wastewater from other enterprises but also reuse recycled water to reap economic benefit. At the same time, the environmental performance is also enhanced, since water scarcity is counteracted by replacing freshwater usage with wastewater and/or recycled water. The benefits of regeneration units in the EIP design will be discussed and the comparison with the case study without regeneration units presented in Part 1 [4] is also considered.

Keywords: Game theory · Nash equilibrium · Single-leader multi-follower game · Mixed-integer programming · Eco-industrial park

1 Introduction

The growing scarcity of freshwater coupled with an increasing demand for water suggests that the need to reuse and recycle water is essential. For this reason, the concept of EIPs was born [1]. In an EIP, the attempt of businesses cooperate with the others is to reduce waste and pollution, efficiently share resources (e.g., materials, energy, water, by-products, and so on). It also helps achieve sustainable development, with the aim to increase economic gains and to reduce the impact of contaminants on the environment. To implement EIP in a sustainable way, companies need methods and optimization tools to design appropriate inter-enterprises exchanges. EIP problems for managing industrial water can be solved by mathematical programming procedures. Furthermore, two scenarios are considered in the literature for designing an EIP: EIP without regeneration units (see, e.g., [3,6]), and EIP with regeneration units (see, e.g., [2,3,5,6]).

© ICST Institute for Computer Sciences, Social Informatics and Telecommunications Engineering 2021
Published by Springer Nature Switzerland AG 2021. All Rights Reserved
P. Cong Vinh and N. Huu Nhan (Eds.): ICTCC 2021, LNICST 408, pp. 162–177, 2021.
https://doi.org/10.1007/978-3-030-92942-8_14

As presented in Part 1 [4], we design and optimize the water exchange networks in EIPs without regeneration units, based on a game theory approach. And this paper is devoted to the optimal design of water exchange networks in EIPs with regeneration units also based on the game theory approach.

The remainder of this paper is organized as follows: Sect. 2 is dedicated the methodology and model formulation which briefly describes the problem addressed in this article and present a water management model in EIPs with regeneration units, based on a single-leader multi-follower model. The reformatting of the EIP modeling problem as a mixed integer linear programming problem is addressed in Sect. 3. The benefits of regeneration units in the EIP design will be discussed and the comparison with the case study without regeneration units presented in Part 1 [4] is considered in Sect. 4. Finally, conclusions and perspectives are presented in Sect. 5.

2   Methodology and Model Formulation

2.1   Problem Statement

Let n, m denote the given number of enterprises and regeneration units; and $I_P := \{1, \ldots, n\}$, $I_R = \{n+1, \ldots, n+m\}$ denote the index set of enterprises and regeneration units, respectively; and 0 denote a sink node. The sink node is a place to store contaminated wastewater. Thus, we define $I = I_P \cup I_R$ and $I_0 = \{0\} \cup I$. Each enterprise has its own pre-defined water input requirement and quality characteristics, as well as the quantity and quality of available output wastewater. For each enterprise, the resource consumption can be freshwater, wastewater from other enterprises, and/or from regeneration units. Indeed, the polluted water from an enterprise can be sent to the sink node, to other enterprises, and/or to regeneration units.

The objective of the model is to determine a network of connections of water streams among them so that both the total freshwater consumption and the annualized operating cost of each enterprise in the park are minimized, while satisfying all process and environmental constraints.

2.2   Minimizing Operating Costs

Each enterprise $i \in I_P$ may receive the wastewater from other enterprises, and/or from regeneration units within the EIP. Nevertheless, for technical constraints on the enterprise i, the contaminant concentration of total flux, delivered by the other enterprises and/or by the regeneration units, cannot exceed a certain maximum value denoted here by $C_{i,\text{in}}$ [ppm]. On the other hand each enterprise $i \in I_P$ generates a mass of contaminant M_i [g/h] due to its own working, that needs to be diluted before exiting the enterprise. To do so, enterprise i should buy an amount of freshwater z_i [T/h] such that, after dilution, the output pollutant concentration is lower than a limit concentration $C_{i,\text{out}}$ [ppm]. Actually considering that enterprise i will optimize his process, it is assumed that each

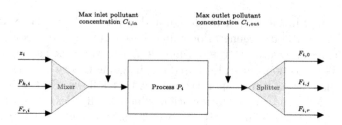

Fig. 1. Process around an enterprise i.

enterprise $i \in I$ will consume the exact amount of the freshwater it needs to attain concentration constraints $C_{i,\text{out}}$, and therefore, its output pollutant concentration will have a concentration equivalent to this constant $C_{i,\text{out}}$. Obviously we have that $C_{i,\text{in}} \leq C_{i,\text{out}}$. This structure is illustrated in Fig. 1.

In addition, each regeneration unit $r \in I_R$ has a given output contaminant concentration, that is $C_{r,\text{out}}$ [ppm]. Moreover, the enterprises can send the wastewater to the regeneration units in order to reduce the contaminant concentration of wastewater. Then, then enterprises will buy and use the recycle water. The process around a regeneration unit is illustrated in Fig. 2.

Fig. 2. Process around a regeneration unit r.

There is an exchange of materials between the enterprises in EIPs. On the other hand, let E be the configuration for the water exchange network in EIPs such that $(i, j) \in E$ then the enterprise i can pump his wastewater to enterprise j. Especially if the enterprise i uses the connection $(i, 0)$, it means that it is discharging polluted water outside the park.

Defining the *stand-alone* and *complete* configuration, respectively, as follows

$$E_{\text{st}} := \{(i, 0) : i \in I_P\}$$
$$E_{\text{max}} := \{(i, j) : i \in I, j \in I_0\} \cup \{(r, k) : r \in I_R, k \in I_P\},$$

thus a valid configuration E must satisfy that $E_{\text{st}} \subset E \subset E_{\text{max}}$. We denote $\mathcal{E}$ as the set of valid configurations for the EIP. In addition, for any $E \in \mathcal{E}$, we denote by $E^c = E_{\text{max}} \backslash E$ the family of connections that are not present in E.

In term of variables, each enterprise $i \in I_P$ sends polluted water to $j \in I_0$, taken into account by variable $F_{i,j}$ [T/h]. In addition, we set $F = (F_{i,j} : (i, j) \in E_{\text{max}})$ to be the full vector of fluxes through the exchange network.

Furthermore, for each enterprise $i \in I$, we denote $F = (F_i, F_{-i}^P, F^R)$ where $F_i = (F_{i,j} : j \in I)$, $F_{-i}^P = (F_{k,j} : k \in I_P \backslash \{i\})$, and $F^R = (F_r : r \in I_R)$, to emphasize the vector of fluxes between enterprise $i \in I_P$. Then, for a fixed network E, the EIP model with regeneration units must satisfy the following constraints:

1. Water mass balance constraint around an enterprise $i \in I_P$:

$$z_i + \sum_{(k,i) \in E} F_{k,i} = \sum_{(i,j) \in E} F_{i,j}. \tag{1}$$

2. Contaminant mass balance constraint around an enterprise $i \in I_P$:

$$M_i + \sum_{(k,i) \in E} C_{k,\text{out}} F_{k,i} = C_{i,\text{out}} \sum_{(i,j) \in E} F_{i,j}, \tag{2}$$

3. Inlet/outlet concentration constraints around an enterprise $i \in I_P$:

$$\sum_{(k,i) \in E} C_{k,\text{out}} F_{k,i} \leq C_{i,\text{in}} \left(z_i + \sum_{(k,i) \in E} F_{k,i} \right). \tag{3}$$

4. Contaminant concentration constraints around a regeneration unit $r \in I_R$:

$$\sum_{(k,r) \in E} C_{k,\text{out}} F_{k,r} \geq C_{r,\text{out}} \sum_{(r,j) \in E} F_{r,j}. \tag{4}$$

5. Mass balance around a regeneration unit $r \in I_r$:

$$\sum_{(k,r) \in E} F_{k,r} = \sum_{(r,j) \in E} F_{r,j}. \tag{5}$$

6. Positivity of fluxes and null fluxes outside the connections: we need all the fluxes to be positive:

$$\begin{cases} F_{i,j} \geq 0, & \forall (i,j) \in E \\ z_i \geq 0, & \forall i \in I_P. \end{cases} \tag{6}$$

Of course, we also put

$$\forall (i,j) \in E^c, \, F_{i,j} = 0, \tag{7}$$

that is, the effective fluxes can only pass through the existing connections in E.

By combining Eqs. (1) and (2) we obtain:

$$M_i + \sum_{(k,i) \in E} C_{k,\text{out}} F_{k,i} = C_{i,\text{out}} \left(z_i + \sum_{(k,i) \in E} F_{k,i} \right), \quad \forall i \in I_p. \tag{8}$$

From Eq. (8) the freshwater consumption of the enterprise $i \in I$ is defined by

$$z_i(F_{-i}) = \frac{1}{C_{i,\text{out}}} \left(M_i + \sum_{(k,i) \in E} (C_{k,\text{out}} - C_{i,\text{out}}) F_{k,i} \right). \tag{9}$$

Thus, each enterprise $i \in I_P$ wants to minimize his operating cost, given by

$$\text{Cost}_i(F_i, F_{-i}^P, F^R, E) = A\Bigg[c \cdot z_i(F_{-i}) + \gamma_{i,0}F_{i,0} + \sum_{r \in I_R} \gamma_{i,r}F_{i,r}$$
$$+ \sum_{r \in I_R} \gamma_{r,i}F_{r,i} \sum_{k \in I_P} \gamma_{k,i}F_{k,i} + \sum_{j \in I_P} \gamma_{i,j}F_{i,j} + \sum_{r \in I_R} \mu_r F_{r,i}^{\psi}\Bigg],$$

where A [h] stands for the annual EIP operating hours, c [\$/T] for the price to buy freshwater, $\gamma_{i,0}$[\$/T] for the price of wastewater discharge, $\gamma_{p,q}$[\$/T] for the cost of sending wastewater from enterprise p to q, and μ_r[\$/T] for the cost of regenerating water. The power $\psi < 1$ in the regenerated water cost term accounts for economy scale, namely the larger the volume of regenerated water, the lower the operating cost of the regeneration unit.

With all these considerations, each enterprise's $i \in I_P$ optimization problem is given by $P_i\left(F_{-i}^P, F^R, E\right)$:

$$\min_{F_i} \text{Cost}_i\left(F_i, F_{-i}^P, F^R, E\right)$$
$$s.t. \begin{cases} \text{Equations (1)-(3)-(6)-(7).} \end{cases} \tag{10}$$

For an operation of the regeneration units F^R and a configuration of the exchange network $E \in \mathcal{E}$, the family of equilibria for F^R and E at the lower-level problem is given by $\text{Eq}(F^R, E)$. Furthermore,

$$\text{Eq}(F^R, E) \iff \forall i \in I_P, \ F_i \text{ solves the problem } P_i\left(F_{-i}^P, F^R, E\right).$$

Remark 1. *If no enterprises and no regeneration units send water to enterprise $i \in I_P$, the amount of freshwater consumed by enterprise i should be*

$$z_i = \frac{M_i}{C_{i,\text{out}}}.$$

Then, the cost of stand-alone configuration is given by

$$\text{STC}_i = A \cdot (c + \gamma_{i,0})\frac{M_i}{C_{i,\text{out}}}.$$

2.3   Minimizing Consumption of Natural Resources

In the model, the EIP authority tries to optimize the total freshwater consumption, and so he wants to minimize the objective function

$$Z(F) = \sum_{i \in I_P} z_i(F_{-i}). \tag{11}$$

The EIP authority must guarantee that each enterprise $i \in I_P$ participating in the EIP reduces its operating cost compared to the stand-alone case, that is,

$$\text{Cost}_i(F_i, F_{-i}^P, F^R, E) \leq \alpha \cdot \text{STC}_i. \tag{12}$$

Now, the optimization problem of the EIP authority is given by

$$\min_{F \in \mathbb{R}^{|E_{\max}|}, E \in \mathcal{E}} Z(F)$$

$$s.t. \begin{cases} \text{Equations (4)-(5)-(6),} \\ F^P \in \text{Eq}(F^R, E), \\ \text{Cost}_i(F_i, F^P_{-i}, F^R, E) \leq \alpha_i \cdot \text{STC}_i, & \forall i \in I_P. \end{cases} \qquad (13)$$

3   Mixed-Integer Programming Reduction

As we can observe the EIP authority problem (13) is mathematical programming with equilibrium constraints, and thus it is difficult to solve. In the following, however, this type of problem will be reformulated as a single mixed-integer programming problem.

3.1   Characterization of Equilibria

In order to characterize the set of the equilibria $\text{Eq}(F^R, E)$ as a system of equalities and inequalities, for each enterprise $i \in I_P$, we define the set

$$E_{i,\text{act}} := \left\{ (i,j) \in E \ : \ \gamma_{i,j} = \gamma_i^* := \min_{(i,k) \in E} \gamma_{i,k} \right\}. \qquad (14)$$

Theorem 2. *For any valid exchange network $E \in \mathcal{E}$ and $F^R \geq 0$, and denoting $S(E)$ by the set*

$$S(F^R, E) = \left\{ F^P \ : \forall i \in I_P, \begin{array}{l} z_i(F_{-i}) + \sum_{(k,i) \in E} F_{k,i} = \sum_{(i,j) \in E} F_{i,j} \\[2mm] \sum_{(k,i) \in E} C_{k,\text{out}} F_{k,i} \leq C_{i,\text{in}} \left(z_i + \sum_{(k,i) \in E} F_{k,i} \right) \\[2mm] F_i \big|_{E^c_{i,\text{act}}} = 0 \\[1mm] F_i \geq 0 \\[1mm] z_i(F_{-i}) \geq 0 \end{array} \right\} \qquad (15)$$

then, one has $S(F^R, E) = \text{Eq}(F^R, E)$. Furthermore, any optimal solution (F, E) of the mathematical programming problem

$$\min_{F \in \mathbb{R}^{|E_{\max}|}, E \in \mathcal{E}} Z(F)$$

$$s.t. \begin{cases} \text{Equations(4)} - (5) - (6), \\ F^P \in S(F^R, E), \\ \text{Cost}_i(F_i, F^P_{-i}, F^R, E) \leq \alpha_i \cdot \text{STC}_i, & \forall i \in I_P. \end{cases} \qquad (16)$$

is an optimal solution of the SLMF problem (13).

For the proof of Theorem 2, we prefer the reader to Appendix A.

3.2   Mixed-Integer Formulation

Constraint $F_i\big|_{E_{i,\text{act}}^c} = 0, \forall i \in I$ depends on the configuration of the water exchange network and is therefore difficult to implement numerical experiments. Thus, we will reduce the single optimization problem (13) to the mixed-integer programming problem.

Now, let $(i,j) \in E_{\max}$, we define

$$C(i,j) := \begin{cases} \{(i,k) \in E_{\max} \; : \; \gamma_{i,k} = \gamma_{i,j}\} & \text{if } i \in I_P \\ \{(i,k) \in E_{\max}\} & \text{if } i \in I_R \end{cases} \tag{17}$$

the *arc class* of (i,j). Moreover, we denote by $\mathcal{C}_i = \{C(i,j) \; : \; (i,j) \in E_{\max}\}$ the set of all arc classes of enterprise i.

If there exists a class $C(i,j) \in \mathcal{C}_i$ satisfying

$$E_{i,\text{act}} \subseteq C(i,j), \tag{18}$$

then it is called the *active class* of E of the enterprise i, and we will denote it as $C_i(E)$.

Now, let $D = \bigcup_{i \in I} \mathcal{C}_i$, the family of all arc classes of enterprises. For each enterprise $i \in I$ and each arc class $C \in \mathcal{C}_i$, we introduce the boolean variable $y = (y_C)_{C \in D} \in \{0,1\}^{|D|}$ in the following way:

$$y_C = \begin{cases} 1 & \text{if } C \text{ is the active class of } i, \\ 0 & \text{otherwise.} \end{cases}$$

From this new boolean variable $y \in \{0,1\}^{|D|}$, we will build a configuration associated to y as

$$E(y) = \left(\bigcup\{C \; : \; y_C = 1\}\right) \cup \{(i,0) \; : \; i \in I_P\} \cup \{(r,j) \in E_{\max} \; : \; r \in I_R\}.$$

Now, let's consider the following Mixed-Integer optimization problem:

$$\min_{F \in \mathbb{R}^{|E_{\max}|}, y \in \{0,1\}^{|D|}} Z(F)$$

$$\text{s.t.} \begin{cases} \text{Equations (1)-(3)-(4)-(5)-(6),} & \\ \sum_{C \in \mathcal{C}_i} y_C = 1, & \forall i \in I_P, \\ \sum_{(i,j) \in C} F_{i,j} \leq K \cdot y_C, & \forall C \in D, \\ \text{Cost}_i(F_i, F_{-i}^P, F^R, E) \leq \alpha_i \cdot \text{STC}_i, & \forall i \in I_P, \end{cases} \tag{19}$$

Here, K is a constant large enough.

Theorem 3. *If (F, E) is an optimal solution of the problem* (16)*, then* (F, y^E) *is an optimal solution of the problem* (19)*, where* $y^E \in \{0, 1\}^{|D|}$ *is given by*

$$y_C^E = \begin{cases} 1 & \text{if } C = C_i(E) \text{ for some } i \in I, \\ 0 & \text{otherwise.} \end{cases}$$

If (F, y) is an optimal solution of the problem (19)*, then $(F, E(y))$ is an optimal solution of the problem* (16)*.*

For the proof of Theorem 3, we prefer the reader to Appendix B.

3.3   Null Class as Exit Option

We can observe that the stand-alone configuration E_{st} is always a feasible configuration for the EIP model. However, with the constraint (12), the problem may become infeasible. Therefore, we need to take into account the possibility of excluding some enterprises from the network when the EIP authority does not ensure to satisfy constraint (12) for such enterprises.

Now, for each enterprise $i \in I$, we introduce a boolean variable $y_{i,\mathrm{null}} \in \{0, 1\}$ such that

$$y_{i,\mathrm{null}} = \begin{cases} 1 & \text{if } i \text{ breaks the constraint } (12), \\ 0 & \text{otherwise.} \end{cases}$$

With this boolean variable, we will add the following constraints to problem (19):

1. For each enterprise $i \in I_P$,

$$y_{i,\mathrm{null}} + \sum_{C \in \mathcal{C}_i} y_C = 1, \tag{20}$$

namely, only one class is active.
2. For each enterprise $i \in I_P$,

$$\sum_{(i,j) \in C(i,0)} F_{i,j} \leq K \cdot (y_{C(i,0)} + y_{i,\mathrm{null}}), \tag{21}$$

$$\sum_{(i,j) \in E_{\max}, j \neq 0} F_{i,j} \leq K \cdot (1 - y_{i,\mathrm{null}}). \tag{22}$$

Constraints (21) and (22) are to ensure that, if the enterprise breaks the constraint, then the enterprise e do not share polluted water with other enterprises and will use the connection $(i, 0)$.
3. For each enterprise $i \in I_P$,

$$\sum_{(k,i) \in E_{\max}} F_{k,i} \leq K \cdot (1 - y_{i,\mathrm{null}}). \tag{23}$$

This constraint is to ensure that, if the enterprise breaks the constraint (12), then no enterprises can send him any polluted water.

4. For each enterprise $i \in I_P$,

$$\mathrm{Cost}_i(F_i, F_{-i}, E(y)) \leq \alpha_i \cdot \mathrm{STC}_i \cdot (1 - y_{i,\mathrm{null}}) + \mathrm{STC}_i \cdot y_{i,\mathrm{null}}. \qquad (24)$$

We denote by $\overline{D} = D \cup \{\mathrm{Null}_i \ : \ i \in I_P\}$, where Null_i is the null class, associated to $y_{i,\mathrm{null}}$, and $D_0 = D \backslash \{C(i,0) : i \in I_P\}$. Denoting

$$\mathrm{STC}_i(y_{i,\mathrm{null}}) := \alpha_i \cdot \mathrm{STC}_i \cdot (1 - y_{i,\mathrm{null}}) + \mathrm{STC}_i \cdot y_{i,\mathrm{null}},$$

With all the foregoing, problem (19) becomes

$$\min_{F \in \mathbb{R}^N, y \in \{0,1\}^{|\overline{D}|}} Z(F)$$

$$s.t. \begin{cases} \text{Equations } (1)\text{-}(3)\text{-}(4)\text{-}(5)\text{-}(6)\text{-}(20)\text{-}(21)\text{-}(22)\text{-}(23), \\ \displaystyle\sum_{(i,j) \in C} F_{i,j} \leq K \cdot y_C, & \forall C \in D_0, \\ \mathrm{Cost}_i(F_i, F_{-i}, E(y)) \leq \mathrm{STC}_i(y_{i,\mathrm{null}}), & \forall i \in I_P. \end{cases} \qquad (25)$$

4   Numerical Experiments

4.1   Case Study

Now, we simulate numerical examples of the model described in Sect. 2. We assume that

$$\gamma_{i,j} = \begin{cases} \delta & \text{if } j \in I_P, \\ 2\delta & \text{if } j \in I_R, \\ \beta & \text{if } j = 0. \end{cases} \qquad (26)$$

From (26), we can observe that each enterprise i has to pay for both water receipt and water deposit. In the case of regeneration units, the enterprises must pay for the operation of the regeneration units. Hence, for each enterprise $i \in I_P$, the set $\mathcal{C}_i = \{C_{i,p}, C_{i,r}, C_{i,0}\}$ where

$$C_{i,p} = \{(i,j) \in E_{\max} \ : \ j \in I_P\}$$
$$C_{i,r} = \{(i,r) \in E_{\max} \ : \ r \in I_R\}$$
$$C_{i,0} = \{(i,0)\}.$$

Now, for each enterprise $i \in I_P$, we introduce four integer variables, $y_{i,p}, y_{i,r} y_{i,0}, y_{i,\mathrm{null}} \in \{0,1\}$ as follows:

- If $y_{i,p} = 1$, it means the connections in $C_{i,p}$ are included in the network.
- If $y_{i,r} = 1$, it means the connections in $C_{i,r}$ are included in the network.
- If $y_{i,0} = 1$, it means the connection $(i,0)$ is the only exit connection for i, and i participates in the EIP.
- If $y_{i,\mathrm{null}} = 1$, it means the connection $(i,0)$ is the only exit connection for i, and i does not participate in the EIP.

Note that only one of these integer variables takes the value 1, i.e., $y_{i,\text{null}} + y_{i,p} + y_{i,r} + y_{i,0} = 1, \forall i \in I$, and in doing so it determines the network E to be implemented and the operation that each enterprise can do within this network. We denote $y \in \{0,1\}^{4n}$ as the vector of all integer variables of all enterprises.

Since the optimization problem (25) can have several solutions and to get a solution with more enterprises involved, we replace $Z(F)$ by

$$Z(F) + \text{Coef} \cdot \sum_{i \in I_P} y_{i,\text{null}}, \tag{27}$$

where $\text{Coef} \geq 0$ is a coefficient to penalize the objective function. Then, the optimization problem (25) becomes:

$$\min_{F,y} \quad Z(F) + \text{Coef} \cdot \sum_{i \in I} y_{i,\text{null}}$$

$$\text{s.t.} \begin{cases} \text{Equations (1)-(3)-(4)-(5)-(6),} \\ y_{i,\text{null}} + y_{i,p} + y_{i,r} + y_{i,0} = 1, & \forall i \in I_P, \\ \sum_{(i,j) \in C_{i,p}} F_{i,j} \leq K \cdot y_{i,p}, & \forall i \in I_P, \\ \sum_{(i,j) \in C_{i,r}} F_{i,j} \leq K \cdot y_{i,r}, & \forall i \in I_P, \\ F_{i,0} \leq K \cdot (y_{i,0} + y_{i,\text{null}}), & \forall i \in I_P, \\ \sum_{(i,j) \in E_{\max}, j \neq 0} F_{i,j} \leq K \cdot (1 - y_{i,\text{null}}), & \forall i \in I_P, \\ \sum_{(k,i) \in E_{\max}} F_{k,i} \leq K \cdot (1 - y_{i,\text{null}}), & \forall i \in I_P, \\ \text{Cost}_i(F_i, F_{-i}, E(y)) \leq \cdot \text{STC}_i(y_{i,\text{null}}), & \forall i \in I_P. \end{cases} \tag{28}$$

We use input data on the scale of an EIP of 15 enterprises. The data of enterprises are given in Table 1 and Table 2 of paper [4]. It is supposed that the EIP operates $A = 1$ h. Additionally, we assume that there are 3 different regeneration units that are distinguished by their ability to regenerate water. The operating parameters of regeneration units are illustrated in Table 6 of the paper [6].

All simulations below performed using Julia Julia v1.0.5 programming language, using CPLEX V12.10.0 as a solver.

4.2   Results and Discussion

The resulting optimized EIP network is shown in Fig. 3, and it corresponds to $\alpha = 0.90$ and $\text{Coef} = 1$. This optimal network provides operating cost of each enterprise and total freshwater consumption that are lower than a stand-alone network as shown in Table 1.

Table 1. Summary of results of the EIP.

Enterprise	Freshwater stand-alone (T/h)	Freshwater in EIP (T/h)	$Cost_i$ stand-alone (MMUSD/hour)	$Cost_i$ in EIP (MMUSD/hour)	% Reduction in $Cost_i$
1	20.00	20.00	7.00	3.00	57.14
2	25.00	25.00	8.75	3.50	60.00
3	30.00	12.50	10.50	7.33	30.15
4	50.00	18.52	17.50	10.31	41.08
5	66.67	0.00	23.33	14.01	39.94
6	8.33	0.00	2.92	1.60	45.07
7	25.00	0.00	8.75	5.90	32.55
8	37.50	0.00	13.12	8.51	35.18
9	100.00	0.00	35.00	28.13	19.62
10	33.33	0.00	11.67	7.46	36.01
11	33.33	0.00	11.67	9.92	14.94
12	25.00	0.00	8.75	7.67	12.38
13	27.27	0.00	9.54	8.19	14.24
14	4.28	0.00	1.50	0.13	91.18
15	10.00	0.00	3.50	1.45	58.55
Total	495.72	76.02	173.50	117.13	32.49

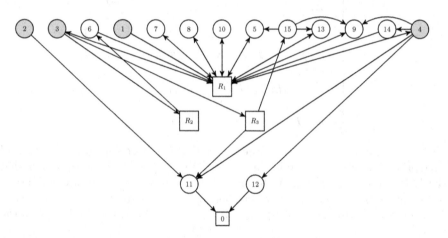

Fig. 3. The optimal configuration in the case $\alpha_i = 0.90$ and Coef $= 1$. Gray nodes are active consuming fresh water.

When enterprises operate in an EIP with an optimal configuration, the benefits of using water regeneration units are obvious. First, total operating cost is reduced compared to the stand-alone case, as expected, from 495.72 (T/h) to 76.02 (T/h), which equates to a reduction of 84.66%. Furthermore, the water demand of enterprises 5, 6, 7, 8, 9, 10, 11, 12, 13, 14, 15 are entirely supplied by other enterprises and/or by regeneration units. Secondly, the EIP authority ensures that each enterprise gets at least a 10% reduction in costs. Enterprise 14 achieves the highest percentage reduction of operating cost corresponding to 91.18% while enterprise 12 has the lowest reduction corresponding to 12.38% with respect to the stand-alone configuration. Total operating cost is reduced compared to the stand alone case, as expected, from 173.50 ($/h) to 117.13 ($/h), which means a decrease of 32.49%.

4.3   Comparison Between the EIP Model Without Regeneration Units and the EIP Model with Regeneration Units

The design and optimization of industrial water networks in eco-industrial parks are studied by formulating and solving SLMF game problems. The SLMF game methodology is suitable on a case study of water management in EIPs without and with regeneration units. It is therefore important to compare the optimal results of both approaches. The criteria we used for comparing the two models are as follows: first, the ability to reduce freshwater consumption; second, the number of enterprises involved into the optimal EIP and last, the ability to reduce the operating costs of each enterprise.

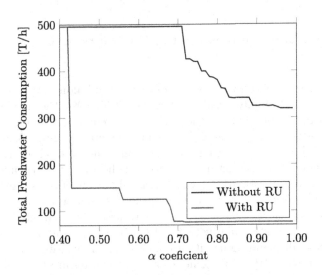

Fig. 4. Sensitivity analysis for $\alpha \in [0.40, 0.99]$ and Coef $= 1$. Total freshwater consumption for both study cases: without and with regeneration units.

First of all and for both models, the total of freshwater consumption and wastewater discharge decreases when α increases as shown in Fig. 4. But the case study of water integration in EIP with regeneration units allows an strong reduction of freshwater consumption compare to the case study of water integration in EIP without regeneration units. In the optimized networks, the EIP with regeneration units achieves a minimum total of freshwater consumption is 76.02

(T/h) corresponding to 84.66% compared to the stand-alone case, while the EIP without regeneration units achieves a minimum total freshwater consumption is 319.04 (T/h) corresponding to 35.64% with respect to the stand-alone configuration.

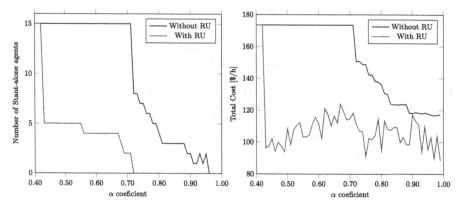

(a) Number of stand-alone enterprises in the park.

(b) Total operating cost in the park.

Fig. 5. Sensitivity analysis for $\alpha \in [0.40, 0.99]$ and Coef = 1.

The number of enterprises which operate stand-alone for both models is shown in Fig. 5a. Roughly speaking, for $\alpha \in [0.40, 0.99]$, the number of enterprises operating stand-alone in the EIP with regeneration units is always less than that of enterprises in the EIP without regeneration units. Moreover, with $\alpha \in [0.43, 0.71]$, the EIP with regeneration units shows its clear superiority on the EIP without regeneration units. Indeed in the EIP with regeneration units the designer can build a park for which not only a reduction of freshwater consumption is achieved compare to EIP without regeneration units results but also the designer can attract the exigent enterprises by guaranteeing a relative gain of more than 29% while with the EIP model without regeneration units only proposes full stand alone situation. Another interesting feature is that if $\alpha \geq 0.72$ then the EIP with regeneration units can ensure that all enterprises will participate in the park while the EIP without regeneration units reach this full involvement only for $\alpha \geq 0.96$. Finally, for $\alpha \leq 0.42$, the optimal solution is the stand-alone configuration for both models, thus 0.42 playing the role of a threshold value for the relative gain.

As observed from Fig. 5b, the total operating cost in the EIP with regeneration units is always less than the one without regeneration units. In the optimized networks, the EIP with regeneration units achieves a minimum total operating cost is 88.58 ($/h) corresponding to 48.94% with respect to the stand-alone configuration, while the EIP without regeneration units achieves a minimum total operating cost is 116.70 (\$/h) corresponding to 32.74% compared to the stand-alone configuration.

5   Conclusion and Perspectives

In this work, we have presented a game theory methodology to optimize the water exchange networks in the EIP. The SLMF game methodology is suitable on a case study of water management in EIP with regeneration units. This study shows that regeneration units yield very significant gains in the EIP design. More precisely, if the enterprises work in the EIP with regeneration units, then the total freshwater consumption and total operating cost in the optimal configuration have been reduced by 84.66% and 32.49%, respectively, while if the enterprises work in the EIP without regeneration units, then the total freshwater consumption and total operating cost in the optimal configuration have been reduced by 34.5% and 31.94%, respectively.

A methodology taking into account in this present work only the single contaminant case is implemented. Thus, we would like to address the EIP design with multi contaminant case in the future. Moreover, the eco-industrial park concept can be extended by sharing not only water but also more resources such as energy or other materials.

Acknowledgments. This research is funded by Nguyen Tat Thanh University, Ho Chi Minh city, Vietnam.

Appendix A

Proof of Theorem 2

We only need to prove the equality $S(F^R, E) = \text{Eq}(F^R, E)$ since the second part of the statement is a direct consequence by replacing the constraint "$F \in \text{Eq}(F^R, E)$" with "$F \in S(F^R, E)$".

First, we show that $S(F^R, E) \subseteq \text{Eq}(F^R, E)$. Indeed, let $F^P \in S(F^R, E)$. Since $E_{i,\text{act}} \subset E$, hence F_i is a feasible point of $P_i(F_{-i}^P, F^R, E)$, for any $i \in I_P$.

Now, let F_i' be another feasible point of $P_i(F_{-i}^P, F^R, E)$, for any $i \in I_P$. Then, $F_i' \geq 0$ and the water mass balance constraint (1) is satisfied. Therefore, one has

$$\text{Cost}_i(F_i', F_{-i}^P, F^R, E) - \text{Cost}_i(F_i, F_{-i}^P, F^R, E) = \sum_{(i,j) \in E} \gamma_{i,j} F_{i,j}' - \gamma_i^* \left(\sum_{(i,j) \in E_{i,\text{act}}} F_{i,j} \right)$$

$$\geq \gamma_i^* \left(\sum_{(i,j) \in E} F_{i,j}' - \sum_{(i,j) \in E_{i,\text{act}}} F_{i,j} \right).$$

Furthermore, the mass balance constraint (1) is satisfied for F_i' and F_i, thus

$$\sum_{(i,j) \in E} F_{i,j}' = z_i(F_{-i}) + \sum_{(k,i) \in E} F_{k,i} = \sum_{(i,j) \in E} F_{i,j} = \sum_{(i,j) \in E_{i,\text{act}}} F_{i,j}.$$

Hence, $\text{Cost}_i(F_i', F_{-i}^P, F^R, E) \geq \text{Cost}_i(F_i, F_{-i}^P, F^R, E)$. Thus, F_i solves $P_i(F_{-i}, E)$, and since this holds for every $i \in I$, we conclude that $F \in \text{Eq}(F^R, E)$.

Now, we show that $\text{Eq}(F^R, E) \subseteq S(F^R, E)$. Indeed, let $F \in \text{Eq}(F^R, E)$, and assume that $F \notin S(F^R, E)$. Since F_i is a feasible point of $P_i(F_{-i}^P, F^R, E)$ for each $i \in I_P$, thus $F \notin S(F^R, E)$ if there exists $i_0 \in I_P$ such that $F_{i_0}|_{E_{i_0,\text{act}}^c} \neq 0$. Hence, there exists $(i_0, j_0) \in E \backslash E_{i_0,\text{act}}$ such that $F_{i_0,j_0} > 0$. Let $(i_0, j_1) \in E_{i,\text{act}}$ and consider the vector

$$F_{i_0,k}' = \begin{cases} F_{i_0,k} & \text{if } k \in I \backslash \{j_0, j_1\}, \\ 0 & \text{if } k = j_0, \\ F_{i_0,j_1} + F_{i_0,j_0} & \text{if } k = j_1. \end{cases}$$

We have that $F_{i_0}' \geq 0$ and also

$$z_i(F_{-i_0}) + \sum_{(k,i_0) \in E} F_{k,i_0} = \sum_{(i_0,j) \in E} F_{i_0,j} = \sum_{(i_0,j) \in E} F_{i_0,j}'.$$

Hence F_{i_0}' is a feasible point of $P_i(F_{-i_0}^P, F^R, E)$. Furthermore, we have that

$$\begin{aligned} \text{Cost}_{i_0}(F_i'0, F_{-i_0}^P, F^R, E) - \text{Cost}_{i_0}(F_{i_0}, F_{-i_0}^P, F^R, E) &= \sum_{(i_0,j) \in E} \gamma_{i_0,j} F_{i_0,j}' - \sum_{(i_0,j) \in E} \gamma_{i_0,j} F_{i_0,j} \\ &= (\gamma_{i_0,j_1} - \gamma_{i_0,j_0}) F_{i_0,j_0} \\ &= (\gamma^* - \gamma_{i_0,j_0}) F_{i_0,j_0} < 0, \end{aligned}$$

since, by construction, $\gamma_{i_0,j_0} > \gamma^*$. This show that F_{i_0} doesn't solve $P_i(F_{-i_0}^P, F^R, E)$, which is a contradiction. Therefore, $F \in S(F^R, E)$.

Appendix B

Proof of Theorem 3

Let (F, E) be an optimal solution of problem (16). Since, for each enterprise $i \in I$, $E_{i,\text{act}} \subseteq C(i,j)$ thus there exists a unique active class $C_i(E)$. Let us define $y^E \in \{0,1\}^{|D|}$ with

$$y_C^E = \begin{cases} 1 & \text{if } C = C_i(E) \text{ for some } i \in I, \\ 0 & \text{otherwise.} \end{cases}$$

Then, for every $i \in I_P$, $\sum_{C \in C_i} y_C^E = 1$. Now, fix a class $C \in D$, and let $i \in I_P$ such that $C \in C_i$. We have that

$$\sum_{(i,j) \in C} F_{i,j} \leq \begin{cases} K = K \cdot y_C^E & \text{if } C = C_i(E), \\ 0 = K \cdot y_C^E & \text{if } C \neq C_i(E), \end{cases}$$

For an enterprise $i \in I_P$, the fact that $E_{i,\text{act}} \subseteq E(y^E)$ lead us to the fact that

$$\text{Cost}_i(F_i, F_{-i}^P, F^R, E(y^E)) = \text{Cost}_i(F_i, F_{-i}^P, F^R, E),$$

and so, the constraint (12) is satisfied. We deduce that (F, y^E) is a feasible point of (19), since all other constraints are directly satisfied given that (F, E) is feasible for problem (16). Therefore, (F, y^E) is an optimal solution of the problem (19).

Now, let (F, y) be an optimal solution of problem (19). For each $i \in I_P$ and let C_i be the unique class in $\mathcal{C}_i$ satisfying $y_{C_i} = 1$. Then, we have

$$E(y)_{i,\text{act}} = C_i \text{ and } \sum_{(i,j) \in E_{\max} \setminus C_i} F_{i,j} \leq K \cdot \sum_{C \in \mathcal{C}_i \setminus \{C_i\}} y_C = 0.$$

We then infer that

$$F_i \big|_{E(y)_{i,\text{act}}^c} = 0.$$

Since this constraint is valid for every active enterprise $i \in I_P$, we can now rewrite the water mass balance constraint in problem (19) as

$$z(F_{-i}) + \sum_{(k,i) \in E(y)} F_{k,i} = \sum_{(i,j) \in E(y)} F_{i,j}, \forall i \in I.$$

We conclude that $(F, E(y))$ is a feasible point of problem (16), thus $(F, E(y))$ is an optimal solution of problem (16).

References

1. Boix, M., Montastruc, L., Azzaro-Pantel, C., Domenech, S.: Optimization methods applied to the design of eco-industrial parks: a literature review. J. Clean. Prod. **87**, 303–317 (2015)
2. Boix, M., Montastruc, L., Pibouleau, L., Azzaro-Pantel, C., Domenech, S.: A multiobjective optimization framework for multicontaminant industrial water network design. J. Environ. Manag. **92**(7), 1802–1808 (2011)
3. Boix, M., Montastruc, L., Pibouleau, L., Azzaro-Pantel, C., Domenech, S.: Industrial water management by multiobjective optimization: from individual to collective solution through eco-industrial parks. J. Clean. Prod. **22**(1), 85–97 (2012)
4. Cao Van, K.: A game theory approach for water exchange in eco-industrial parks: part 1 - a case study without regeneration units. Accepted to 7th EAI International Conference on Nature of Computation and Communication, p. 15pp (2021)
5. Montastruc, L., Boix, M., Pibouleau, L., Azzaro-Pantel, C., Domenech, S.: On the flexibility of an eco-industrial park (EIP) for managing industrial water. J. Clean. Prod. **43**, 1–11 (2013)
6. Ramos, M., Boix, M., Aussel, D., Montastruc, L., Domenech, S.: Water integration in eco-industrial parks using a multi-leader-follower approach. Comput. Chem. Eng. **87**, 190–207 (2016)

Numerical Solution of Robin-Dirichlet Problem for a Nonlinear Wave Equation with Memory Term

Le Thi Mai Thanh[1,2,3], Tran Trinh Manh Dung[2,3], and Nguyen Huu Nhan[1(✉)]

[1] Nguyen Tat Thanh University, 300A Nguyen Tat Thanh Street,
Dist. 4, Ho Chi Minh City, Vietnam
{ltmthanh,nhnhan}@ntt.edu.vn
[2] Faculty of Mathematics and Computer Science, University of Science,
Ho Chi Minh City, 227 Nguyen Van Cu Street, Dist. 5, Ho Chi Minh City, Vietnam
[3] Vietnam National University, Ho Chi Minh City, Vo Truong Toan Street,
Linh Trung Ward, Thu Duc City, Vietnam

Abstract. In this paper, we propose a numerical procedure to find the approximate solution of initial-boundary value problem for a nonlinear wave equation with memory term. First, the local existence and uniqueness is proved by the linear approximating method and the Faedo-Galerkin method. Next, a numerical scheme is constructed by the finite-difference method and the standard arguments of ordinary differential equation. Finally, an example is given to simulate the numerical results of the proposed algorithm.

Keywords: Wave equation · Faedo-Galerkin method ·
Finite-difference method · Robin-Dirichlet condition · Memory term

AMS subject classification: 35L05 · 35L15 · 35L20 · 35L55 · 35L70

1 Introduction

In the present paper, we are concerned with the following initial-boundary value problem with memory term

$$u_{tt} - \frac{\partial}{\partial x}\left(\mu(x,t)u_x\right) + \int_0^t g(t-s)\frac{\partial}{\partial x}\left(\bar{\mu}(x,s)u_x(x,s)\right)ds = f(x,t,u,u_t),$$
$$0 < x < 1, 0 < t < T, \qquad (1)$$

$$u_x(0,t) - u(0,t) = u(1,t) = 0, \qquad (2)$$

$$u(x,0) = \tilde{u}_0(x), u_t(x,0) = \tilde{u}_1(x), \qquad (3)$$

where μ, $\bar{\mu}$, g, f, $\tilde{u}_0$ and $\tilde{u}_1$ are given functions which satisfy conditions specified later.

P. Cong Vinh and N. Huu Nhan (Eds.): ICTCC 2021, LNICST 408, pp. 178–191, 2021.
https://doi.org/10.1007/978-3-030-92942-8_15

As is well known, the integral term in the Eq. (1) is the memory term responsible for the viscoelastic property. The wave equation with memory term (or so called viscoelastic term) is arised in studies about viscoelastic materials, which stands for a capacity of storage and dissipation of mechanical energy. The dynamic properties of viscoelastic materials are very importance and interest because they appear in many applied sciences, for more literature on this topic, one can see [1,3,4] and references therein.

During few past decades, the mathematical models similar to the problem (1)–(3) have been extensively studied by many researchers. Indeed, for multi-dimension spatial problems, numerous works are dedicated to the investigation of the general form of viscoelastic wave equation obtained by

$$u_{tt} - \Delta u + \int_0^t g(t-s)\Delta u(x,s)ds - \lambda\Delta u_t + \gamma h\left(u_t\right) = \mathcal{F}(u), \qquad (4)$$

see [2,5,6,9,12,17] for details and their references. When $\lambda = 0$, $\gamma = 1$, $h \equiv u_t$, $\mathcal{F} \equiv u\left|u\right|^{p-1}$, Kafin and Messaoudi [5] investigated the following Cauchy problem on $\mathbb{R}^n$, in which some conditions have been put on the kernel g to get the finite-time blow up of solution of the corresponding problem. Recently, Li and He [6] have studied the Eq. (4) for a bounded domain $\Omega \subset \mathbb{R}^n$ and $\lambda = \gamma = 1$, $h \equiv u_t$, $\mathcal{F} \equiv u\left|u\right|^{p-2}$, i.e., they considered the following nonlinear viscoelastic wave equation

$$u_{tt} - \Delta u + \int_0^t g(t-s)\Delta u(x,s)ds - \Delta u_t + u_t = u\left|u\right|^{p-2}, \qquad (5)$$

where $\Omega \subset \mathbb{R}^n$ is a bounded domain with smooth boundary $\partial\Omega$. Then the global existence, general decay and blow-up properties of solutions for the above model were proved. For one-dimension spatial problems, some interesting results of existence, exponential decay and asymptotic expansion have been investigated in different contexts. For example, in [19], Ngoc et al. proved a local existence of the wave equation with nonlinear convolution as follows

$$u_{tt} - \lambda u_{xxt} - \frac{\partial}{\partial x}\left[\mu_1\left(x,t,u(x,t),\|u(t)\|^2,\|u_x(t)\|^2\right)u_x\right]$$
$$+ \int_0^t g(t-s)\frac{\partial}{\partial x}\left[\mu_2\left(x,s,u(x,s),\|u(s)\|^2,\|u_x(s)\|^2\right)u_x(x,s)\right]ds \qquad (6)$$
$$= F\left(x,t,u,u_x,u_t,\|u(t)\|^2,\|u_x(t)\|^2\right), 0 < x < 1, 0 < t < T,$$

associated with Robin-Dirichlet boundary conditions (2) and initial conditions (3), where $\lambda > 0$ is a constant, and μ_1, μ_2, g, f are given functions which satisfy some certain conditions. Moreover, the authors established an approximation of weak solution by the asymptotic expansion method, satisfying an estimation with N order. Before, Long and his colleagues [7] have discussed the equation (6) in case of $\lambda = 0$, $\mu_1 = \mu_2 = 1$, $F \equiv f(x,t,u) - \left|u_t\right|^{q-2}u_t$, associated with mixed nonhomogeneous boundary conditions and initial conditions (3).

Then they proved the global existence and exponential decay of weak solution of the corresponding model, in addition an example was given to present the numerical approximating solution of the problem.

Although a numerous of works studying of solution properties to viscoelastic problems were published, however, it seems that few publications of this type coupled with numerical results are considered. For some last decades, there has been much effort to develop different numerical methods for the solution of partial differential equations with viscoelastic term, see for instance as in [7,13, 16,19,20] and their references. In [19], Quynh et al. considered a specific form of (6) with $\lambda = 0$, $\mu_1 = \mu_2 = 1$, $F \equiv f(x,t,u) - \alpha |u_t|^{p-2} u_t$ together with Robin boundary conditions

$$u_{tt} - u_{xx} + \alpha |u_t|^{p-2} u_t + \int_0^t g(t-s)u_{xx}(x,s)ds$$
$$= f(x,t,u), 0 < x < 1, 0 < t < T. \tag{7}$$

The authors established the existence of N-order recurrent sequence associated with the Eq. (7) and proved that this sequence converges to the unique solution of the Eq. (7). Moreover, by using finite-difference formulas, they constructed an algorithm in order to find numerical solutions given by the 2-order iterative scheme (when $N = 2$). Before, in [7,13], the authors have also used finite-difference approximation to establish numerical solution for wave equations with viscoelastic term or with integral boundary conditions, respectively. Beside finite-difference method, several methods were developed to solve partial differential equations such as cubic spline approximation or finite-element method. For example, Mohanty and Gopal [8] applied the high accuracy cubic spline finite difference method for one-dimensional nonlinear wave equation, in which the application of the proposed method to telegraphic equation and wave equation in polar coordinates was considered, and their stability was also analyzed. Furthermore, the comparisons between the numerical results of proposed high accuracy cubic spline finite difference method and the corresponding second order accuracy cubic spline method were discussed. In [20], Saedpanah studied the initial-boundary value problem with memory

$$u''(t) - Au(t) + \int_0^t K(t-s)Au(s)ds = f(t), t \in (0,T),$$
$$u(0) = u^0, \ u'(0) = u^1, \tag{8}$$

(u' is used for $\dfrac{du}{dt}$), associated with a homogeneous Dirichlet boundary condition or with a mixed Dirichlet-Neumann boundary condition, where A is a self-adjoint, positive definite, second order elliptic linear operator on a certain separable Hilbert space. The kernel K is considered to be either smooth (exponential), or no worse than weakly singular. By using Picard's iteration, the existence and uniqueness of the spatial local and global Galerkin approximation of the problem (8) were proved. Then, spatial finite element scheme of the problem was constructed and optimal order a priori estimates were proved by the energy method. Finally, the author gave a numerical example to illustrate the order of convergence of the spatial finite element discretization to a concrete problem with smooth convolution kernel. In our paper, in Sect. 2, we first introduce some notations and inequalities of compact imbeddings in Hilbert spaces. Next, in Sect. 3, we present the theorem of existence and uniqueness of the problem (8) of which their proofs are the same as in [14, 15, 19]. In Sect. 4, we construct an algorithm in order to find the approximate solution of the problem (1)–(3) by using the finite-difference formulas, and an example is also given to illustrate the exact solution and the finite-difference approximate solution. Finally, in Sect. 5, we summarize the obtained results in our paper.

2   Preliminaries

In this section, we present some notations and materials in order to present main results.

Let $\Omega = (0, 1)$, $Q_T = (0, 1) \times (0, T)$ and we define the scalar product in L^2 by

$$\langle u, \ v \rangle = \int_0^1 u(x)v(x)dx,$$

and the corresponding norm $\|\cdot\|$, i.e., $\|u\|^2 = \langle u, \ u \rangle$. Let us denote the standard function spaces by $C^m(\overline{\Omega})$, $L^P = L^P(\Omega)$ and $H^m = H^m(\Omega)$ for $1 \leq p \leq \infty$ and $m \in \mathbb{N}$. Also, we denote that $\|\cdot\|_X$ is a norm in a certain Banach space X, and $L^p(0, T; X)$, $1 \leq p \leq \infty$, is the Banach space of real functions $u : (0, T) \to X$ measurable with the corresponding norm $\|\cdot\|_{L^p(0,T;X)}$ defined by

$$\|u\|_{L^p(0,T;X)} = \left(\int_0^T \|u(t)\|_X^p \, dt \right)^{1/p} < \infty \ for 1 \leq p < \infty,$$

and

$$\|u\|_{L^\infty(0,T;X)} = ess \sup_{0<t<T} \|u(t)\|_X \ \text{for } p = \infty.$$

Let $u(t)$, $u'(t) = u_t(t) = \dot{u}(t)$, $u''(t) = u_{tt}(t) = \ddot{u}(t)$, $u_x(t) = \nabla u(t)$, $u_{xx}(t) = \Delta u(t)$, denote $u(x,t)$, $\dfrac{\partial u}{\partial t}(x,t)$, $\dfrac{\partial^2 u}{\partial t^2}(x,t)$, $\dfrac{\partial u}{\partial x}(x,t)$, $\dfrac{\partial^2 u}{\partial x^2}(x,t)$, respectively.

For $f \in C^k([0,1] \times \mathbb{R}_+ \times \mathbb{R}^2)$, $f = f(x, t, y_1, y_2)$, we put $D_1 f = \dfrac{\partial f}{\partial x}$, $D_2 f = \dfrac{\partial f}{\partial t}$, $D_{2+i} f = \dfrac{\partial f}{\partial y_i}$, $i = 1, 2$ and $D^\alpha f = D_1^{\alpha_1}...D_4^{\alpha_4} f$; $\alpha = (\alpha_1, ..., \alpha_4) \in \mathbb{Z}_+^4$, $|\alpha| = \alpha_1 + ... + \alpha_4 = k$, $D^{(0,...,0)} f = D^{(0)} f = f$.

Put

$$V = \{v \in H^1 : \ v(1) = 0\}, \tag{9}$$

which is a closed subspace of H^1.

We consider a symmetric bilinear form $a(\cdot, \cdot)$ defined by

$$a(u, v) = \int_0^1 u_x(x)v_x(x)dx + u(0)v(0), \text{ with } u, v \in V, \tag{10}$$

and corresponding norm $\|v\|_V = \sqrt{a(v, v)}$.

Then, it is easy to prove the following inequalities (see [15])

(i) $\|v\|_{C^0(\bar{\Omega})} \leq \|v_x\| \leq \|v\|_V$, for all $v \in V$,
(ii) $|a(u, v)| \leq 2\|u\|_V \|v\|_V$, for all $u, v \in V$,
(iii) $a(v, v) \geq \|v_x\|^2$, for all $v \in V$.

3   Existence and Uniqueness

In this section, we consider the local existence and uniqueness of the problem (1)–(3). By using the linear approximate method and the Faedo-Galerkin method, we shall prove that there exists a recurrent sequence converging to the weak solution of (1)–(3) defined as below.

Definition 1. *A function u is called a weak solution of initial-boundary value problem (1)–(3) if $u \in W_T = \{u \in L^\infty(0, T; H^2 \cap V) : u' \in L^\infty(0, T; V), u'' \in L^\infty(0, T; L^2)\}$ and satisfies the variational equation*

$$\langle u''(t), v \rangle + A(t; u(t), v) - \int_0^t g(t - s)\bar{A}(s; u(s), v)ds$$
$$= \langle f(t, u(t), u'(t)), v \rangle, \forall v \in V, \tag{11}$$

together with initial conditions

$$u(0) = \tilde{u}_0, u'(0) = \tilde{u}_1, \tag{12}$$

where

$$A(t; u, v) = \langle \mu(t)u_x, v_x \rangle + \mu(0, t)u(0)v(0), \tag{13}$$
$$\bar{A}(t; u, v) = \langle \bar{\mu}(t)u_x, v_x \rangle + \bar{\mu}(0, t)u(0)v(0), \tag{14}$$

Consider a fixed constant $T^* > 0$, we make the following assumptions:

(H_1) $(\tilde{u}_0, \tilde{u}_1) \in (V \cap H^2) \times V, \tilde{u}_{0x}(0) - \tilde{u}_0(0) = 0$;
(H_2) $g \in H^1(0, T^*)$;
(H_3) $f \in C^1([0, 1] \times [0, T^*] \times \mathbb{R}^2)$;
(H_4) $\mu \in C^2([0, 1] \times [0, T^*])$ and there exists a constant $\mu_0 > 0$ such that $\mu(x, t) \geq \mu_0$ for all $(x, t) \in [0, 1] \times [0, T^*]$;
(H_5) $\bar{\mu} \in C^1([0, 1] \times [0, T^*])$.

For every $T \in (0, T^*]$ and $M > 0$, we put

$$
\begin{cases}
W(M,T) = \{v \in L^\infty(0,T; V \cap H^2) : v_t \in L^\infty(0,T;V), \ v_{tt} \in L^2(Q_T), \\
\quad \text{with } \max\{\|v\|_{L^\infty(0,T;V \cap H^2)}, \|v_t\|_{L^\infty(0,T;V)}, \|v_{tt}\|_{L^2(Q_T)}\} \le M\}, \\
W_1(M,T) = \{v \in W(M,T) : v_{tt} \in L^\infty(0,T; L^2)\},
\end{cases}
$$

in which $Q_T = \Omega \times (0,T)$.

We consider the recurrent sequence $\{u_m\}$ satisfying the first term $u_0 \equiv \tilde{u}_0$, and suppose that

$$u_{m-1} \in W_1(M,T), \tag{15}$$

then we find $u_m \in W_1(M,T)$ $(m \ge 1)$ satisfying the linear variational problem

$$
\begin{cases}
\langle u_m''(t), w \rangle + A(t; u_m(t), w) + \displaystyle\int_0^t g(t-s)\bar{A}(s; u_m(s), w)ds \\
\qquad\qquad = \langle F_m(t), w \rangle, \ \forall w \in V, \\
u_m(0) = \tilde{u}_0, u_m'(0) = \tilde{u}_1,
\end{cases} \tag{16}
$$

where

$$F_m(x,t) = f[u_{m-1}](x,t) = f\left(x, t, u_{m-1}, u_{m-1}'\right). \tag{17}$$

In the next part, we present two theorems that confirm the existence and uniqueness of solutions of the problem (1)–(3). Actually, the first theorem (Theorem 1) claims the existence of the recurrent sequence defined by (15)–(17), after that the second theorem (Theorem 2) shows that the recurrent sequence converges to the unique weak solution of the problem (1)–(3). The proofs of these theorems can be proved similarly to the ones given in [14, 15, 19].

Theorem 1. *Suppose that $(H_1) - (H_5)$ hold. Then there exist positive constants M and T such that, with $u_0 \equiv \tilde{u}_0$, there exists the recurrent sequence $\{u_m\}$ defined by (15)–(17).*

Using the result of Theorem 1 and the compact imbedding theorems, we can establish the existence and uniqueness of weak solution of the problem (1)–(3) which is given the following theorem.

Theorem 2. *Suppose that $(H_1) - (H_5)$ hold. Then, the problem (1)–(3) has a unique weak solution $u \in W_1(M,T)$, where the constants $M > 0$ and $T > 0$ are suitably chosen.*

Moreover, the recurrent sequence $\{u_m\}$ defined by (15)–(17) strongly converges to u in the Banach space

$$W_1(T) = \{v \in L^\infty(0,T;V) : v' \in L^\infty(0,T; L^2)\},$$

and the following estimate is confirmed

$$\|u_m - u\|_{W_1(T)} \le C_T k_T^m, \text{ for all } m \in \mathbb{N}, \tag{18}$$

where $k_T \in [0,1)$ and C_T is a positive constant which depends on $T, f, g, \tilde{u}_0, \tilde{u}_1$ and k_T.

4   Numerical Results

Consider the following problem:

$$\begin{cases} u_{tt} - \dfrac{\partial}{\partial x}\left(\mu(x,t)u_x\right) + \displaystyle\int_0^t g(t-s)\dfrac{\partial}{\partial x}\left(\bar{\mu}(x,s)u_x(x,s)\right)ds \\ \qquad\qquad = f(x,t,u,u_t), 0 < x < 1, 0 < t < T, \\ u_x(0,t) - u(0,t) = u(1,t) = 0, \\ u(x,0) = \tilde{u}_0(x), u_t(x,0) = \tilde{u}_1(x), \end{cases} \tag{19}$$

where the functions $\mu(x,t), \bar{\mu}(x,t), g(t), f, \tilde{u}_0$ and $\tilde{u}_1$ are defined by

$$\begin{cases} \mu(x,t) = 1 + (1+x)e^{-\mu_0 t}, \bar{\mu}(x,t) = 1 + (1+x)e^{-\bar{\mu}_0 t}, g(t) = e^{-g_0 t}, \\ f(x,t,u,u_t) = -u_t^3 + |u|^3 u + F(x,t), \\ \tilde{u}_0(x) = (1-x)e^{2x}, \tilde{u}_1(x) = -(1-x)e^{2x}, \\ F(x,t) = (1-x)e^{2x}\left[1 - (1-x)^2 e^{4x} - (1-x)^3 e^{6x}\right] \\ \qquad + e^{2x-t}\left[4 - \left(1 - 6x - 4x^2\right)e^{-\mu_0 t}\right] - \dfrac{4}{g_0-1}e^{2x}\left(e^{-t} - e^{-g_0 t}\right) \\ \qquad + \dfrac{1}{g_0 - \bar{\mu}_0 - 1}\left(1 - 6x - 4x^2\right)e^{2x}\left(e^{-(\bar{\mu}_0+1)t} - e^{-g_0 t}\right), \\ g_0 = 2, \mu_0 = \bar{\mu}_0 = 1. \end{cases} \tag{20}$$

The exact solution of the problem (19), with $\mu(x,t)$, $\bar{\mu}(x,t)$, $g(t)$, f, $\tilde{u}_0$ and $\tilde{u}_1$ defined in (20) respectively, is the function u_{ex} given by

$$u_{ex}(x,t) = (1-x)e^{2x-t}. \tag{21}$$

In order to find the numerical solution of the problem (19), we use the difference given by [18] (pages 36 and 43) to approximate the second-order spatial derivatives and then transfer (19) to the following system of first-order ordinary differential equations with the unknowns $u_i(t) \equiv u(x_i,t)$, $v_i(t) = u_i'(t)$

$$\begin{cases} u_i'(t) = v_i(t), i = \overline{1,N}, \\ v_1'(t) = \dfrac{1}{h^2}\left[\alpha_1^*(t)u_1(t) + \alpha_2(t)u_2(t)\right] \\ \qquad - \dfrac{1}{h^2}\displaystyle\int_0^t g(t-s)\left[\bar{\alpha}_1^*(s)u_1(s) + \bar{\alpha}_2(t)u_2(s)\right]ds + f_1\left(t, u_1(t), v_1(t)\right), \\ v_i'(t) = \dfrac{1}{h^2}\left[\alpha_i(t)u_{i-1}(t) + \beta_i(t)u_i(t) + \alpha_{i+1}(t)u_{i+1}(t)\right] \\ \qquad - \dfrac{1}{h^2}\displaystyle\int_0^t g(t-s)\left[\bar{\alpha}_i(s)u_{i-1}(s) + \bar{\beta}_i(s)u_i(s) + \bar{\alpha}_{i+1}(s)u_{i+1}(s)\right]ds \\ \qquad + f_i\left(t, u_i(t), v_i(t)\right), i = \overline{2, N-1}, \\ v_N'(t) = \dfrac{1}{h^2}\left[\alpha_N(t)u_{N-1}(t) + \beta_N(t)u_N(t)\right] \\ \qquad - \dfrac{1}{h^2}\displaystyle\int_0^t g(t-s)\left[\bar{\alpha}_N(s)u_{N-1}(s) + \bar{\beta}_N(s)u_N(s)\right]ds + f_N\left(t, u_N(t), v_N(t)\right), \\ u_i(0) = \tilde{u}_0(x_i), v_i(0) = \tilde{u}_1(x_i), i = \overline{1,N}, \end{cases} \tag{22}$$

where $x_i = ih$, $h = \dfrac{1}{N+1}$, $i = \overline{0, N+1}$ and

$$\alpha_i(t) = \mu_i(t) = \mu(x_i, t),$$
$$\beta_i(t) = -\alpha_i(t) - \alpha_{i+1}(t),$$
$$\alpha_1^*(t) = -\frac{h}{1+h}\alpha_1(t) - \alpha_2(t),$$
$$\bar{\alpha}_i(t) = \bar{\mu}_i(t) = \bar{\mu}(x_i, t), \tag{23}$$
$$\bar{\beta}_i(t) = -\bar{\alpha}_i(t) - \bar{\alpha}_{i+1}(t),$$
$$\bar{\alpha}_1^*(t) = -\frac{h}{1+h}\bar{\alpha}_1(t) - \bar{\alpha}_2(t),$$
$$f_i(t, u_i(t), v_i(t)) = f(x_i, t, u_i(t), v_i(t)) = -v_i^3(t) + |u_i(t)|^3 u_i(t) + F(x_i, t), i = \overline{1, N}.$$

Then, the system (22) is equivalent to

$$\begin{cases} X'(t) = \hat{B}(t)X(t) - \displaystyle\int_0^t g(t-s)\hat{A}(s)X(s)ds + \mathcal{F}(t, X(t)), \\ X(0) = X_0, \end{cases} \tag{24}$$

where

$$\begin{cases} X(t) = (u_0(t), \cdots, u_N(t), v_0(t), \cdots, v_N(t))^T \in \mathfrak{M}_{2N\times 1}, \\ \mathcal{F}(t, X(t)) = (0, \cdots, 0, \mathcal{F}_1(t, X(t)), \cdots, \mathcal{F}_N(t, X(t)))^T \in \mathfrak{M}_{2N\times 1}, \\ \mathcal{F}_i(t, X(t)) = f_i(t, u_i(t), v_i(t)) = -v_i^3(t) + |u_i(t)|^3 u_i(t) + F(x_i, t), i = \overline{1, N}, \\ X_0 = (\tilde{u}_0(x_1), \cdots, \tilde{u}_0(x_N), \tilde{u}_1(x_1), \cdots, \tilde{u}_1(x_N))^T \in \mathfrak{M}_{2N\times 1}, \end{cases} \tag{25}$$

$$\hat{A}(t) = \begin{bmatrix} O & O \\ \frac{1}{h^2}\bar{A}(t) & O \end{bmatrix} \in \mathfrak{M}_{2N}, \hat{B}(t) = \begin{bmatrix} O & E \\ \frac{1}{h^2}A(t) & O \end{bmatrix} \in \mathfrak{M}_{2N}, \tag{26}$$

$$E = \begin{bmatrix} 1 & 0 & \cdots & 0 \\ 0 & 1 & \ddots & \vdots \\ \vdots & \ddots & \ddots & 0 \\ 0 & \cdots & 0 & 1 \end{bmatrix} \in \mathfrak{M}_N, O = O_{N\times N} \in \mathfrak{M}_N,$$

$$A(t) = \begin{bmatrix} \alpha_1^*(t) & \alpha_2(t) & 0 & \cdots & & 0 \\ \alpha_2(t) & \beta_2(t) & \alpha_3(t) & 0 & \cdots & 0 \\ 0 & \ddots & \ddots & \ddots & \ddots & \vdots \\ \vdots & 0 & \ddots & \ddots & \ddots & 0 \\ \vdots & \vdots & \ddots & \alpha_{N-1}(t) & \beta_{N-1}(t) & \alpha_N(t) \\ 0 & 0 & \cdots & 0 & \alpha_N(t) & \beta_N(t) \end{bmatrix} \in \mathfrak{M}_N, \tag{27}$$

$$\bar{A}(t) = \begin{bmatrix} \bar{a}_1^*(t) & \bar{a}_2(t) & 0 & \cdots & \cdots & 0 \\ \bar{a}_2(t) & \bar{\beta}_2(t) & \bar{a}_3(t) & 0 & \cdots & 0 \\ 0 & \ddots & \ddots & \ddots & & \vdots \\ \vdots & 0 & \ddots & \ddots & & 0 \\ \vdots & \vdots & \ddots & \bar{a}_{N-1}(t) & \bar{\beta}_{N-1}(t) & \bar{a}_N(t) \\ 0 & 0 & \cdots & 0 & \bar{a}_N(t) & \bar{\beta}_N(t) \end{bmatrix} \in \mathfrak{M}_N. \tag{28}$$

The nonlinear differential system (24) is solved by using the following linear recursive scheme generated by the nonlinear term $\mathcal{F}_i(t, X(t)) = f_i(t, u_i(t), v_i(t)) = -v_i^3(t) + |u_i(t)|^3 u_i(t) + F(x_i, t)$:

$$\begin{cases} \dfrac{dX^{(m)}}{dt}(t) = \hat{B}(t)X^{(m)}(t) - \displaystyle\int_0^t g(t-s)\hat{A}(s)X^{(m)}(s)ds + \mathcal{F}^{(m)}(t), \\ X^{(m)}(0) = X_0, \end{cases} \tag{29}$$

where

$$\begin{cases} X^{(m)}(t) = \left(u_1^{(m)}(t), \cdots, u_N^{(m)}(t), v_1^{(m)}(t), \cdots, v_N^{(m)}(t)\right)^T \in \mathfrak{M}_{2N\times 1}, \\ \mathcal{F}^{(m)}(t) \equiv \mathcal{F}\left(t, X^{(m-1)}(t)\right) \\ \qquad = \left(0, \cdots, 0, \mathcal{F}_1(t, X^{(m-1)}(t)), \cdots, \mathcal{F}_N(t, X^{(m-1)}(t))\right)^T \in \mathfrak{M}_{2N\times 1}, \\ \mathcal{F}_i(t, X^{(m-1)}(t)) = f_i\left(t, u_i^{(m-1)}(t), v_i^{(m-1)}(t)\right) \\ \qquad = -\left(v_i^{(m-1)}(t)\right)^3 + \left|u_i^{(m-1)}(t)\right|^3 u_i^{(m-1)}(t) + F(x_i, t), i = \overline{1, N}, \\ X_0 = (\tilde{u}_0(x_1), \cdots, \tilde{u}_0(x_N), \tilde{u}_1(x_1), \cdots, \tilde{u}_1(x_N))^T \in \mathfrak{M}_{2N\times 1}. \end{cases} \tag{30}$$

In order to find the numerical solution of the problem (29), we will approximate $\dfrac{dX^{(m)}}{dt}(t_j)$ as follows

$$\frac{dX^{(m)}}{dt}(t_j) \approx \frac{X_{j+1}^{(m)} - X_j^{(m)}}{\Delta t}, \tag{31}$$

$$X_j^{(m)} = X^{(m)}(t_j), t_j = j\Delta t, \Delta t = \frac{T}{M}, j = \overline{0, M}.$$

Therefore

$$\begin{cases} \dfrac{X_{j+1}^{(m)} - X_j^{(m)}}{\Delta t} = \hat{B}_j X_j^{(m)} - \displaystyle\int_0^{t_j} g(t_j - s)\hat{A}(s)X^{(m)}(s)ds \\ \qquad\qquad\qquad + \mathcal{F}^{(m)}(t_j), j = \overline{0, M-1}, \\ X_0^{(m)} = X_0. \end{cases} \tag{32}$$

On the other hand, we use the trapezoidal formula to approximate $\int_0^{t_j} g(t_j - s)\hat{A}(s)X^{(m)}(s)ds$ as follows

$$
\int_0^{t_j} g(t_j - s)\hat{A}(s)X^{(m)}(s)ds \approx \Delta t \left[\frac{g_j \hat{A}_0 X_0^{(m)} + g_0 \hat{A}_j X_j^{(m)}}{2} + \sum_{\nu=1}^{j-1} g_{j-\nu} \hat{A}_\nu X_\nu^{(m)} \right]
$$

$$
= \Delta t \left[\frac{g_j \hat{A}_0 X_0 + g_0 \hat{A}_j X_j^{(m)}}{2} + \sum_{\nu=1}^{j-1} g_{j-\nu} \hat{A}_\nu X_\nu^{(m)} \right],
$$

$$(33)$$

where $g_j = g(t_j)$.

Hence

$$
\begin{cases}
X_{j+1}^{(m)} = \left(\mathbb{I} + \Delta t \hat{B}_j - \frac{1}{2} (\Delta t)^2 g_0 \hat{A}_j \right) X_j^{(m)} \\
\qquad - (\Delta t)^2 \left[\frac{g_j \hat{A}_0 X_0}{2} + \sum_{\nu=1}^{j-1} g_{j-\nu} \hat{A}_\nu X_\nu^{(m)} \right] + \Delta t \mathcal{F}^{(m)}(t_j), j = \overline{1, M-1}, \\
X_0^{(m)} = X_0,
\end{cases}
$$

$$(34)$$

where $\mathbb{I}$ the identity matrix of size $2N$.

We rewrite (34) in the form

$$
\begin{cases}
X_1^{(m)} = \left(\mathbb{I} + \Delta t \hat{B}_0 \right) X_0 + \Delta t \mathcal{F}^{(m)}(0) \equiv \mathcal{F}_0 \left[X_0 \right], \\
X_2^{(m)} = \left(\mathbb{I} + \Delta t \hat{B}_1 - \frac{1}{2} (\Delta t)^2 g_0 \hat{A}_1 \right) X_1^{(m)} - \frac{1}{2} (\Delta t)^2 g_1 \hat{A}_0 X_0 + \Delta t \mathcal{F}^{(m)}(t_1) \\
\qquad \equiv \mathcal{F}_1 \left[X_0, X_1^{(m)} \right], \\
X_{j+1}^{(m)} = \left(\mathbb{I} + \Delta t \hat{B}_j - \frac{1}{2} (\Delta t)^2 g_0 \hat{A}_j \right) X_j^{(m)} \\
\qquad - (\Delta t)^2 \left[\frac{g_j \hat{A}_0 X_0}{2} + \sum_{\nu=1}^{j-1} g_{j-\nu} \hat{A}_\nu X_\nu^{(m)} \right] + \Delta t \mathcal{F}^{(m)}(t_j) \\
\qquad \equiv \mathcal{F}_j \left[X_0, X_1^{(m)}, X_2^{(m)}, \cdots, X_j^{(m)} \right], j = \overline{2, M-1},
\end{cases}
$$

$$(35)$$

in which

$$
\begin{cases}
\mathcal{F}_0\left[X_0\right] = \left(\mathbb{I} + \Delta t \hat{B}_0\right) X_0 + \Delta t \mathcal{F}^{(m)}(0), \\
\mathcal{F}_1\left[X_0, X_1^{(m)}\right] = \left(\mathbb{I} + \Delta t \hat{B}_1 - \frac{1}{2}(\Delta t)^2 g_0 \hat{A}_1\right) X_1^{(m)} - \frac{1}{2}(\Delta t)^2 g_1 \hat{A}_0 X_0 + \Delta t \mathcal{F}^{(m)}(t_1), \\
\mathcal{F}_j\left[X_0, X_1^{(m)}, X_2^{(m)}, \cdots, X_j^{(m)}\right] = \left(\mathbb{I} + \Delta t \hat{B}_j - \frac{1}{2}(\Delta t)^2 g_0 \hat{A}_j\right) X_j^{(m)} \\
\qquad\qquad - (\Delta t)^2 \left[\dfrac{g_j \hat{A}_0 X_0}{2} + \displaystyle\sum_{\nu=1}^{j-1} g_{j-\nu} \hat{A}_\nu X_\nu^{(m)}\right] \\
\qquad\qquad + \Delta t \mathcal{F}^{(m)}(t_j), j = \overline{2, M-1}, \\
\mathcal{F}^{(m)}(0) \equiv \mathcal{F}(0, X_0) = (0, \cdots, 0, \mathcal{F}_1(0, X_0), \cdots, \mathcal{F}_N(0, X_0))^T \in \mathfrak{M}_{2N \times 1}, \\
\mathcal{F}_i(0, X_0) = f_i\left(0, \tilde{u}_0(x_i), \tilde{u}_1(x_i)\right) = -\tilde{u}_1^3(x_i) + |\tilde{u}_0(x_i)|^3\, \tilde{u}_0(x_i) + F(x_i, 0), i = \overline{1, N}.
\end{cases}
$$
(36)

With the positive integer N, M fixed, we find $\left(X_1^{(m)}, X_2^{(m)}, \cdots, X_M^{(m)}\right)$ by (35)–(36) such that

$$
\max_{1 \le j \le M} \left|X_j^{(m)} - X_j^{(m-1)}\right|_1 < \varepsilon = 10^{-3},
$$
(37)

where $|\cdot|_1$ is the norm in the space $\mathbb{R}^{2N}$ given as below

$$
\left|X_j^{(m)} - X_j^{(m-1)}\right|_1 = \sum_{i=1}^{N}\left(\left|u_i^{(m)}(t_j) - u_i^{(m-1)}(t_j)\right| + \left|v_i^{(m)}(t_j) - v_i^{(m-1)}(t_j)\right|\right),
$$
(38)

with

$$
X_j^{(m)} = X^{(m)}(t_j) = \left(u_1^{(m)}(t_j), \cdots, u_N^{(m)}(t_j), v_1^{(m)}(t_j), \cdots, v_N^{(m)}(t_j)\right)^T \in \mathfrak{M}_{2N \times 1}.
$$
(39)

If (37) holds then $X_j^{(m)}$ is chosen as follows $X_j^{(m)} \equiv X_j = (u_1(t_j), \cdots, u_N(t_j), v_1(t_j), \cdots, v_N(t_j))^T$, and then the following error is obtained

$$
E_{N,M}(u) = \max_{1 \le j \le M} \max_{1 \le i \le N} |u_{ex}(x_i, t_j) - U_i(t_j)|.
$$
(40)

For details, Table 1 presents the errors $E_{N,M}(u)$ of the approximate solution $u^{(m)}(x, t)$ of (19) defined by (35)–(36) and the exact solution $u_{ex}(x, t)$ defined by (21) with respect to the values of N and M as below

Table 1. Errors of the approximate solution $u^{(m)}(x,t)$ and the exact solution $u_{ex}(x,t)$

N	M	$E_{N,M}(u)$
5	25	0.105613828788385
10	100	0.036310154878634
15	225	0.020252536096917
20	400	0.013818817721289
25	625	0.010475400779207
30	900	0.008349131471897

Obviously, the errors are decreasing when the values of M and N are increasing.

With the functions given as in (20) and $0 \leq x \leq 1$, $0 \leq t \leq 1.5$ and mesh of $N = 30$ and $M = 900$, Fig. 1 illustrates the surface of the exact solution $u_{ex}(x,t)$ of the problem (19) below.

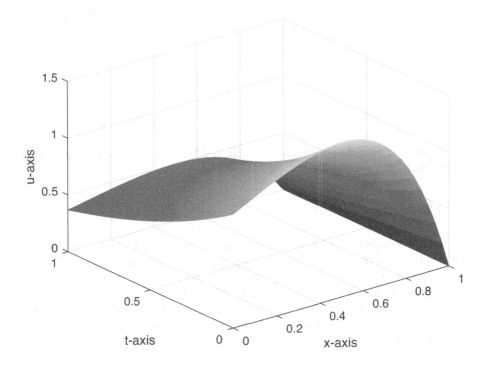

Fig. 1. The picture of the exact solution u_{ex} of (19).

With the functions given as in (20) and mesh of $N = 30$ and $M = 900$, Fig. 2 illustrates the surface of the approximate solution of $u^{(m)}(x,t)$ to the problem (19) with defined by the algorithm (35) and (36).

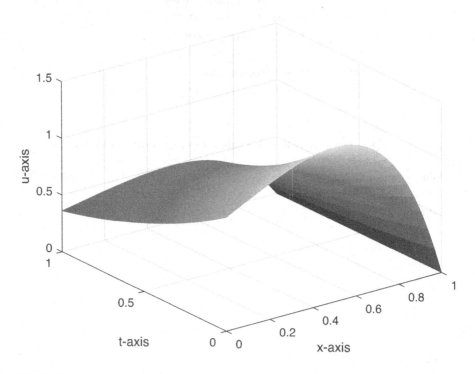

Fig. 2. The picture of the approximate solution of $u^{(m)}(x,t)$ to the problem (19) defined by the algorithm (35) and (36).

5   Conclusion

This paper is concerned with an initial-boundary value problem for nolinear wave equation with memory term. The results of existence and uniqueness of solutions to the problem are established by the linear approximate technique and the Faedo-Galerkin method, in which their proofs are the same as in [14, 15, 19]. An algorithm in order to find the approximate solution of the problem is constucted by the finite-difference formulas, and an example is also given to illustrate the exact solution and the finite-difference approximate solution.

References

1. Duvaut, G., Lions, J.L.: Inequalities in Mechanics and Physics, 1st edn. Springer-Verlag, Berlin Heidelberg (1976)

2. Hao, J., Wei, H.: Blow-up and global existence for solution of quasilinear viscoelastic wave equation with strong damping and source term. Bound. Value Probl. **2017**(1), 1–12 (2017). https://doi.org/10.1186/s13661-017-0796-7

3. Ijaz, N., Bhatti, M., Zeeshan, A.: Heat transfer analysis in magnetohydrodynamic flow of solid particles in non-Newtonian Ree-Eyring fluid due to peristaltic wave in a channel. Therm. Sci. **23**, 1017–1026 (2019)

4. Iqbal, S.A., Sajid, M., Mahmood, K., Naveed, M., Khan, M.Y.: An iterative approach to viscoelastic boundary-layer flows with heat source/sink and thermal radiation. Therm. Sci. **24**, 1275–1284 (2020)

5. Kafini, M., Messaoudi, S.A.: A blow-up result in a Cauchy viscoelastic problem. Appl. Math. Lett. **21**, 549–553 (2008)

6. Li, Q., He, L.: General decay and blow-up of solutions for a nonlinear viscoelastic wave equation with strong damping. Bound. Value Probl. **2018**(1), 1–22 (2018). https://doi.org/10.1186/s13661-018-1072-1

7. Long, N.T., Dinh, A.P.N., Truong, L.X.: Existence and decay of solutions of a nonlinear viscoelastic problem with a mixed nonhomogeneous condition. Numer. Funct. Anal. Optim. **29**(11–12), 1363–1393 (2008)

8. Mohanty, R.K., Gopal, V.: High accuracy cubic spline finite difference approximation for the solution of one-space dimensional nonlinear wave equations. Appl. Math. Comput. **218**, 4234–4244 (2011)

9. Messaoudi, S.A.: General decay of solutions of a viscoelastic equation. J. Math. Anal. Appl. **341**, 1457–1467 (2008)

10. Messaoudi, S.A.: General decay of the solution energy in a viscoelastic equation with a nonlinear source. Nonlinear Anal. TMA. **69**, 2589–2598 (2008)

11. Mustafa, M.I.: Optimal decay rates for the viscoelastic wave equation. Math. Methods Appl. Sci. **41**, 192–204 (2018)

12. Mustafa, M.I.: General decay result for nonlinear viscoelastic equations. J. Math. Anal. Appl. **457**, 134–152 (2018)

13. Ngoc, L.T.P., Triet, N.A., Ngoc Dinh, A.P., Long, N.T.: Existence and exponential decay of solutions for a wave equation with integral nonlocal boundary conditions of memory type, Numer. Funct. Anal. Optim. **38** 1173–1207 (2017)

14. Ngoc, L.T.P., Quynh, D.T.N., Long, N.T.: Linear approximation and asymptotic expansion associated to the Robin-Dirichlet problem for a Kirchhoff-Carrier equation with a viscoelastic term. Kyungpook Math. J. **59**, 735–769 (2019)

15. Nhan, N.H., Ngoc, L.T.P., Thuyet, T.M., Long, N.T.: A Robin-Dirichlet problem for a nonlinear wave equation with the source term containing a nonlinear integral. Lith. Math. J. **57**, 80–108 (2017)

16. Oruç, Ö.: Two meshless methods based on local radial basis function and barycentric rational interpolation for solving 2D viscoelastic wave equation. Comput. Math. Appl. **79**, 3272–3288 (2020)

17. Park, J.Y., Park, S.H.: General decay for quasilinear viscoelastic equations with nonlinear weak damping. J. Math. Phys. **50**, 083505 (2009)

18. Pinder, G.F.: Numerical Methods for Solving Partial Differential Equations: A Comprehensive Introduction for Scientists and Engineers, Wiley and Sons, Hoboken (2018)

19. Quynh, D.T.N., Nam, B.D., Thanh, L.T.M., Dung, T.T.M., Nhan, N.H.: High-order iterative scheme for a viscoelastic wave equation and numerical results. Math. Probl. Eng. **2021**, 27 (2021)

20. Saedpanah, F.: Existence and convergence of Galerkin approximation for second order hyperbolic equations with memory term, Numer. Methods Partial. Differ. Equ. **32**, 548–563 (2016)

Stable Random Vector and Gaussian Copula for Stock Market Data

Truc Giang Vo Thi$^{(\boxtimes)}$ (ID)

Tien Giang University, Long Xuyen, Tien Giang, Vietnam
vothitrucgiang@tgu.edu.vn

Abstract. A lot of real-world data sets have heavy tailed distribution, while the calculations for these distributions in multidimensional case are complex. The paper shows a method to investigate data of multivariate heavy-tailed distributions. We show that for any given number $\alpha \in (0; 2]$, each Gaussian copula is also the copula of an α-stable random vector. Simultaneously, every random vector is α-stable if its marginals are α-stable and its copula is a Gaussian copula. The result is used to build up a formula representing density functions of α-stable random vectors with Gaussian copula. Adopting a new tool, the paper points out that in most of cases, vectors of daily returns in stock market data have multivariate stable distributions with Gaussian copulas and we propose a new method to choose an investment portfolio by computing the distribution of linear combination of components of stable random vector. Dataset of 4 stocks on HOSE was analyzed.

Keywords: Stable distribution · Gaussian copula · Stock market

1 Introduction

Until the 1970's, most of the statistical analysis methods were developed under normality assumptions, mainly for mathematical convenience. In applications, however, normality is only a poor approximation of reality. In particular, normal distributions do not allow heavy-tailed, which are so common, especially in finance and risk management studies [1,7,9,15,18,20]. Arising as solutions to central limit problems, stable distributions are natural heavy tailed extensions of normal distributions and have attracted a lot of attention [9,12,23,24].

While the univariate stable distributions are now mostly accessible by several methods to estimate stable parameters and reliable programs to compute stable densities, cumulative distribution functions, and quantiles for stable random variables [5,10,23,25], the use of the heavy-tailed models in practice has been restricted by the lack of the tools for multivariate stable distributions.

The main challenge of dealing with multivariate data with heavy-tailed distributions is that of ambiguous dependence between coordinates of a random vector. Whilst the dependence can be completely determined by covariance matrix

P. Cong Vinh and N. Huu Nhan (Eds.): ICTCC 2021, LNICST 408, pp. 192–208, 2021.
https://doi.org/10.1007/978-3-030-92942-8_16

for the case of multinormal data, the covariance matrix does not exist for heavy-tailed data. Fortunately, the problem can be solved by the tool of copula. The term copula was first introduced by Sklar [19], but was not of great interest until recent years. Copula functions describe the dependence structure connecting random variables, giving an opportunity to separate the dependence structure and marginal distributions.

Another way of parameterizing multivariate stable distributions is to use the well known univariate stable distribution results about one-dimensional projections of random vectors. However, in practice, this approach faces with challenging computational problems which have not been generally solved for multivariate stable distributions. The problems are caused by the complexity of the possible distributions with an uncountable set of parameters. In recent years, computations are more accessible for elliptically contoured stable distributions [28] which are scale mixtures of multivariate normal distributions. The tools for the very special class of stable distributions were applied in several empirical studies [8,12].

Although the method is available only for a narrow subclass of symmetric multivariate stable distributions, that approach stimulates researchers to create similar tools for other subclasses of general stable multidimensional distributions. We developed a new method for investigating the data of multivariate distributions with heavy tails, trying to decrease the complexity of stable copulas downwards to a more practicable case of Gaussian copulas. This method was applied for daily return data on Nasdaq [16]. We continue to use it for daily return data on HOSE in Vietnam.

The paper is organized as follows. Section 2 presents some auxiliary results on copulas and stable distributions. In Sect. 3 we give the main results of multidimensional stable distributions with Gaussian copulas, demonstrating that Gaussian copulas are also those of some multivariate stable distributions and a random vector is stable if it has Gaussian copula and all its marginals are stable. In the last section we formulate the density function of a stable random vector with Gaussian copula, which can be practically computed. Then the results are applied for study of stock market data of Vietnam.

2   Copulas and Stable Random Vectors

Given a random vector $\mathbf{X} = (X_1, ..., X_d)^T$ taking values in Euclidean space $\mathbb{R}^d$, its cumulative distribution function (cdf hereafter) and probability density function (pdf hereafter) are denoted by $F_{\mathbf{X}}$ and $f_{\mathbf{X}}$, respectively. The coordinates $X_1, ..., X_d$ are called marginals, simultaneously $F_{X_1}, ..., F_{X_d}$ and $f_{X_1}, ..., f_{X_d}$ are called marginal cdf's and marginal pdf's of $\mathbf{X}$, respectively.

Let a increasing function $G : \mathbb{R} \to [0, 1]$, the *generalized inverse* of G is

$$G^{\leftarrow}(y) = inf\{x : G(x) \geq y\}, y \in [0, 1]$$

If G is cdf of continous variable, $G^{\leftarrow}(y) = G^{-1}(y)$.

2.1   Copulas

Copulas are tools for modelling dependence of a random vector with any marginal distributions. In 1959, copula was firstly introduced by Abe Sklar. In 1999, Embrechts, McNeil, Straumann applied copulas in finance. In 2006, copulas were used as a risk management tool in insurance and bank. Up to now, copulas have been used in many fields as sea storm (Corbella and Stretch 2013), risk evaluation of droughts (Zhang et al. 2013).

Let a random vector $\mathbf{X} = (X_1, ..., X_d)$, a d-dimensional *copula* (or d-copula) of $\mathbf{X}$ is the function $C_{\mathbf{X}} : [0, 1]^d \to [0, 1]$ given by

$$C_{\mathbf{X}}(u_1, ..., u_d) = F_{\mathbf{X}}(F_{X_1}^{\leftarrow}(u_1), ..., F_{X_d}^{\leftarrow}(u_d)) \tag{1}$$

where $F_{\mathbf{X}}$ is the cdf, and $F_{X_1}^{\leftarrow}, ..., F_{X_d}^{\leftarrow}$ are generalized inverse functions of marginal distribution functions of $\mathbf{X}$ [27].

The famous Sklar's Theorem [19] confirms the relationship

$$F_{\mathbf{X}}(x_1, ..., x_d) = C_{\mathbf{X}}(F_{X_1}(x_1), ..., F_{X_d}(x_d)) \tag{2}$$

for $x_1, ..., x_d \in \bar{\mathbb{R}} = [-\infty, +\infty]$.

The copula was simply the joint distribution function of random variables with uniform marginals. To manipulate copulas, specific copula types have been introduced, and can be divided into two groups: explicit copulas (Archimedean copulas) and implicit copulas (Gaussian copula,...). The Gaussian copula is the most popular one in applications. It is simply derived from the correlation matrix Σ and mean vector $\boldsymbol{\mu}$ of a multivariate Gaussian distribution function (without loss of generality, we can assume $\boldsymbol{\mu} = \mathbf{0}$) and is given by the following formula:

$$C(u_1, ..., u_d) = \int_{-\infty}^{\phi^{-1}(u_1)} ... \int_{-\infty}^{\phi^{-1}(u_d)} \frac{exp\left[\frac{-1}{2}(\mathbf{x} - \boldsymbol{\mu})^T \Sigma^{-1}(\mathbf{x} - \boldsymbol{\mu})\right]}{\sqrt{(2\pi)^d|\Sigma|}} dx_d...dx_1 \tag{3}$$

$u_i \in [0, 1], i = 1, 2, ..., d$, $\mathbf{x} = (x_1, ..., x_d)^T$, ϕ is cdf of univariable Gaussian distribution. Moreover, the calculation of Gaussian copula is available on computer softwares, such as R software.

One nice property of copulas is that for strictly monotone transformations of the random variables, copulas are either invariant. In general, the copula functions are invariant under strictly increasing transformations. Especially, the following proposition given by Embrechts et al. [1] provides an useful tool for getting the main results of our study.

Proposition 1. *Let $(X_1, ..., X_d)^T$ be a vector of continuous random variables with copula C. If $T_1, ..., T_d : \mathbb{R} \to \mathbb{R}$ are strictly increasing on $RanX_1, ..., RanX_n$, resp, then $(T_1(X_1), ..., T_d(X_d))^T$ also has copula C.*

2.2   Stable Random Vectors

A random variable X is said to have a *stable distribution* if for any positive numbers A, B, there is a positive number C and a real number D such that

$$AX_1 + BX_2 \overset{D}{=} CX + D \qquad (4)$$

where X_1 and X_2 are independent copies of X.

For any stable random variable X, there is a number $\alpha \in (0, 2]$ such that the numbers A, B, C in (4) satisfy the following formula $A^\alpha + B^\alpha = C^\alpha$. The number α is called the index of stability or characteristic exponent. The stable random variable X with index α is called α-stable.

The probability densities of α-stable random variables exist and are continuous but, with a few exceptions, they are not known in closed form [22]. Characteristic function is a useful tool for studying these variables. In particular, *characteristic function* of an α-stable random variable X has the form

$$\hat{X}(u) = \mathbb{E}(e^{iuX}) = \begin{cases} exp(-\gamma^\alpha |u|^\alpha [1 - i\beta(tan\frac{\pi\alpha}{2})sign(u)] + i\delta u) & \alpha \neq 1, \\ exp(-\gamma|u|[1 + i\beta(\frac{2}{\pi})sign(u)ln|u|] + i\delta u) & \alpha = 1 \end{cases} \qquad (5)$$

where $\beta \in [-1, 1], \gamma \in (0, +\infty), \delta \in (-\infty, +\infty)$ and

$$sign(u) = \begin{cases} 1 & if \quad u > 0, \\ 0 & if \quad u = 0, \\ -1 & if \quad u < 0 \end{cases}$$

The parameters α, β, γ and δ uniquely determine the distribution of X, the symbol $X \sim S(\alpha, \beta, \gamma, \delta)$ can be used to refer to that situation. Usually, α is called the stable index, meanwhile β, γ and δ are named as the skewness, the scale and the location parameters of X, respectively.

Next, some basic linear properties of α-stable random variables are used in paper [17].

Proposition 2. *If $X \sim S(\alpha, \beta, \gamma, \delta)$, then*

$$aX + b \sim \begin{cases} S(\alpha, \beta sign(a), |a|\gamma, a\delta + b) & \alpha \neq 1, \\ S(\alpha, \beta sign(a), |a|\gamma, a\delta + b - \beta\gamma\frac{2}{\pi}aln|a|) & \alpha = 1 \end{cases} \qquad (6)$$

$a \neq 0, b \in R.$

Proposition 3. *If $X_1 \sim S(\alpha, \beta_1, \gamma_1, \delta_1)$ and $X_2 \sim S(\alpha, \beta_2, \gamma_2, \delta_2)$ are independent then*

$$X_1 + X_2 \sim S(\alpha, \beta, \gamma, \delta) \qquad (7)$$

where $\beta = \frac{\beta_1\gamma_1^\alpha + \beta_2\gamma_2^\alpha}{\gamma_1^\alpha + \gamma_2^\alpha}, \gamma^\alpha = \gamma_1^\alpha + \gamma_2^\alpha, \delta = \delta_1 + \delta_2.$

Similar to a α-stable random variable in $\mathbb{R}$, stable random vector is expanded in $\mathbb{R}^d$. Characteristic functions are still used to describe the stable distribution in multivariate case, but they are more complicated and determined by three paramaters: stable index α, vecotr μ^0 and spectral measure Λ. Moreover, any linear combination of the components of $\mathbf{X}$ of the type $Y = \sum_{k=1}^{d} b_k X_k$ is an α-stable radom variable. An important property of stable distribution family is the stability of the linear combinations of its coordinates. It completely determines $\mathbf{X}$, as it can be seen from the next theorem.

Theorem 1. *Let $\mathbf{X}$ be an α-stable random vector, every linear combination $\mathbf{u}^T \mathbf{X}$ has the stable distribution $S(\alpha, \beta(\mathbf{u}), \gamma(\mathbf{u}), \delta(\mathbf{u}))$, then the parameter functions are written in the following form:*

$$\gamma(\mathbf{u}) = \left(\int_{\mathbb{S}_d} |\mathbf{u}^T \mathbf{s}| \Lambda(ds) \right)^{1/\alpha}$$

$$\beta(\mathbf{u}) = \gamma(\mathbf{u})^{-\alpha} \int_{\mathbb{S}_d} |\mathbf{u}^T \mathbf{s}| sign(\mathbf{u}^T \mathbf{s}) \Lambda(ds)$$

$$\delta(\mathbf{u}) = \begin{cases} \mathbf{u}^T \mu^0 & , \alpha \neq 1 \\ \mathbf{u}^T \mu^0 - \frac{2}{\pi} \int_{\mathbb{S}_d} |\mathbf{u}^T \mathbf{s}|.log(\mathbf{u}^T \mathbf{s}) \Lambda(ds) & , \alpha = 1 \end{cases}$$

3   Stable Random Vector and Gaussian Copula

This section shows main contents of the paper. From that, we apply for actual data in next section.

Lemma 1. *Let Y and Z be continuous random variables with pdf's f_Y and f_Z which are positive on the images $ran(Y)$ and $ran(Z)$ of Y and Z. Then there exists a strictly increasing function $g : ran(Y) \rightarrow ran(Z)$ such that the random variable $g \circ Y : \Omega \rightarrow ran(Z)$ has the same distribution as Z. Moreover, the function g has positive derivative g'.*

Proof. By assumption, the pdf's f_Y and f_Z are positive, therefore the cdf's F_Y and F_Z are strictly increasing on $ran(Y)$ and $ran(Z)$, respectively. Then the function $g : ran(Y) \rightarrow ran(Z)$ defined by $g(u) = F_Z^{-1}(F_Y(u))$ is welldetermined as a strictly increasing function. Besides, for each $u \in ran(Z)$,

$$g'(u) = \frac{f_Y(u)}{f_Z(g(u))}$$

which is a positive function. The identity implies

$$f_Z(g(u))g'(u)du = f_Y(u)du$$

that yields

$$F_Z(g(u)) = \int_{-\infty}^{g(u)} f_Z(g(u))g'(u)du = \int_{-\infty}^{g(u)} f_Y(u)du = F_Y(u) \qquad (8)$$

On the other hand, for every $t \in \mathbb{R}$,

$$F_{g \circ Y}(t) = \mathbb{P}(\omega : g(Y(\omega)) \leq t) = \mathbb{P}(\omega : Y(\omega) \leq g^{-1}(t)) = F_Y(g^{-1}(t))$$

Compared the above with (8), putting $t = g(u)$ implies

$$F_Z(t) = F_{g \circ Y}(t)$$

This confirms the two random variables Z and $g \circ Y$ have the same distribution. The lemma is proved. $\square$

Although the lemma seems to be quite trivial and simple, it can be useful in practical application. Namely in applied statistics, sometimes normalizing transformations that turn a given data set to a new form with normal distribution need to be used. One of those normalizing transformation for continuously distributed data is proposed in the following immediate consequence of the lemma.

Corollary 1. *Let X be continuous random variable with pdf f_X which is positive on $ran(X)$. Then there exists a strictly increasing function $g : \mathbb{R} \to \mathbb{R}$ such that the random variable $g \circ X$ has normal distribution.*

Corollary 2. *Let X be a normal random variable and a positive number $\alpha \in (0, 2]$. Then there exists a strictly increasing function $g : \mathbb{R} \to \mathbb{R}$ such that the random variable $g \circ X$ is α-stable.*

Corollary 1, 2 is obtained from Lemma 1 when Z is the normal random variable, α-stable, respectively.

It is evident that all marginals of a stable random vector are stable random variables. The inverse statement is not true, a random vector with all stable marginals is not always stable. However, as it is confirmed in the next lemma, the inverse statement is valid if those marginals are independent.

Lemma 2. *Let $\alpha \in (0, 2]$ and a random vector $\boldsymbol{U} = (U_1, ..., U_d)^T$ be given. Supposed that the marginals $U_1, ..., U_d$ are independent α-stable random variables, then $\boldsymbol{U}$ is an α-stable random vector.*

Proof. Let two random vectors $\mathbf{U}^1 = (U_1^1, ..., U_d^1)$ and $\mathbf{U}^2 = (U_1^2, ..., U_d^2)$ which independent and identically distributed with $\mathbf{U} = (U_1, ..., U_d)$. With two positive real numbers A, B, we will prove existence of a positive real number C and vector $\mathbf{D} \in \mathbb{R}^d$ such that $A\mathbf{U}^1 + B\mathbf{U}^2 \overset{D}{=} C\mathbf{U} + \mathbf{D}$.

Because $\mathbf{U}^1, \mathbf{U}^2$ independent and identically distributed, $U_1, ..., U_d$ are independent α-stable random variables, thus every pair of U_k^1, U_k^2 are independent and identically distributed α-stable, $k = 1, ..., d$. Consider two positive numbers A, B, with every pair of U_k^1, U_k^2, there exist positive number $C = (A^\alpha + B^\alpha)^{1/\alpha}$ and D_k such that

$$AU_k^1 + BU_k^2 \overset{D}{=} CU_k + D_k \tag{9}$$

This means every $t_k \in \mathbb{R}$, $F_{AU_k^1+BU_k^2}(t_k) = F_{CU_k+D_k}(t_k)$, $k = 1, .., d$. Based on the assumption, $U_1, ..., U_d$ are independent, we have $U_1^1, ..., U_d^1$ and $U_1^2, ..., U_d^2$ also are independent. Combine this with (9), we have

$$
\begin{aligned}
F_{A\mathbf{U}^1+B\mathbf{U}^2}(t_1, ..., t_d) &= F_{AU_1^1+BU_1^2}(t_1) \times ... \times F_{AU_d^1+BU_d^2}(t_d) \\
&= F_{CU_1+D_1}(t_1) \times ... \times F_{CU_d+D_d}(t_d) \\
&= F_{C\mathbf{U}+\mathbf{D}}(t_1, ..., t_d)
\end{aligned}
$$

So, $A\mathbf{U}^1 + B\mathbf{U}^2 \overset{D}{=} C\mathbf{U} + \mathbf{D}$. This ensures that $\mathbf{U}^1 = (U_1^1, ..., U_d^1)$ is an α-stable random vector. $\square$

Proposition 4. *Let* $\boldsymbol{X} = (X_1, ..., X_d)$ *be an d-dimensional normally distributed random vector, where* $X_1, ..., X_d$ *are independence. Then, every* $\alpha \in (0, 2]$, *there exists a strictly increasing function* $g : \mathbb{R} \to \mathbb{R}$ *such that the random vector* $\boldsymbol{Y} = (g \circ X_1, ..., g \circ X_d)$ *is* α-*stable.*

Proof. Because $\mathbf{X} = (X_1, ..., X_d)$ is normal random vector, its components $X_1, ..., X_d$ have normal distribution. By Corollary 2, every $k = 1, ..., d$, there exists a strictly increasing function $g : \mathbb{R} \to \mathbb{R}$ such that the random variable $g \circ X_k : \Omega \to \mathbb{R}$ is α-stable. Otherwise, $X_1, ..., X_d$ are independent and g is strictly increasing, so $g \circ X_1, ..., g \circ X_d$ are also independent. Then, Corollary 2 ensures the random vector $\mathbf{Y} = (g \circ X_1, ..., g \circ X_d)$ is α-stable. $\square$

The Lemma 2 shows that the independence is a strong condition that guarantees the stability of a random vector with stable marginals. However, for every 2-stable random vector, it is always possible to rotate the space axes to get a new basis, in which the random vector has independent marginals. Namely, the standard Cholesky decomposition theorem immediately implies the following.

Lemma 3. *Let* $\boldsymbol{X}$ *be an d-dimensional normally distributed random vector with expextation* $\boldsymbol{\mu}$ *and positive defined covariance matrix* Σ, $\boldsymbol{X} \sim N_d(\boldsymbol{\mu}, \Sigma)$. *Then there exists an orthogonal* $d \times d$ *matrix* $A = (a_{ij})$ *such that the r.v.* $\boldsymbol{Y} = A\boldsymbol{X}$ *has independent normally distributed marginals, where* $A : \mathbb{R}^d \to \mathbb{R}^d$ *is the linear transformation defined by* $AX = AX^T$.

Proof. We have matrix Σ positive defined, so there exists orthogonal $d \times d$ matrix $A = (a_{ij})$ such that $A^{-1}\Sigma A$ is diagonal matrix.

Set $\mathbf{Z} = (Z_1, ..., Z_d) = A\mathbf{X}$. Because $\mathbf{X} = (X_1, ..., X_d)$ is an d-dimensional normally distributed random vector, we have $X_1, ..., X_d$ are normal variables, so every random variable $Z_k = \sum_{j=1}^d a_{kj}X_j$ has normal distribution.

Overwise,

$$
\begin{aligned}
(\mathbf{x} - \boldsymbol{\mu})^T \Sigma^{-1}(\mathbf{x} - \boldsymbol{\mu}) &= (A^{-1}\mathbf{z} - A^{-1}A\boldsymbol{\mu})^T \Sigma^{-1}(A^{-1}\mathbf{z} - A^{-1}A\boldsymbol{\mu}) \\
&= (\mathbf{z} - A\boldsymbol{\mu})^T (A^{-1})^T \Sigma^{-1}(A^{-1})(\mathbf{z} - A\boldsymbol{\mu}) \\
&= (\mathbf{z} - A\boldsymbol{\mu})^T A\Sigma^{-1}A^{-1}(\mathbf{z} - A\boldsymbol{\mu}) \\
&= (\mathbf{z} - \boldsymbol{\mu})^T (A^{-1}\Sigma A)^{-1}(\mathbf{z} - \boldsymbol{\mu})
\end{aligned}
$$

This implies $\mathbf{Z}$ has normal distribution with covariance matrix $A^{-1}\Sigma A$ which is diagonal matrix. Therefore, $\mathbf{Z}$'s components $Z_1, ..., Z_d$ do not correlate, means independent. $\qquad\square$

Proposition 5. *Let $\mathbf{X} = (X_1, ..., X_d)$ be an d-dimensional normally distributed random vector and $\alpha \in (0, 2]$. Then there exists an orthogonal $d \times d$ matrix $A = (a_{ij})$ and a strictly increasing function $g : \mathbb{R} \to \mathbb{R}$ such that $\mathbf{Y} = (g \circ \sum_{j=1}^{d} a_{1j}X_j, ..., g \circ \sum_{j=1}^{d} a_{dj}X_j)$ is α-stable.*

Proof. With $\mathbf{X}$ is normal random vector, by Lemma 3, there exists an orthogonal $d \times d$ matrix $A = (a_{ij})$ such that $\mathbf{Z} = A\mathbf{X}$ has independent normally distributed marginals. Let $\alpha \in (0, 2]$, by Proposition 4, there exists the strictly increasing function $g : \mathbb{R} \to \mathbb{R}$ such that random vector $\mathbf{Y} = (g \circ \sum_{j=1}^{d} a_{1j}X_j, ..., g \circ \sum_{j=1}^{d} a_{dj}X_j)$ is α-stable. Proposition is proven. $\qquad\square$

We are now ready to state the main result.

Theorem 2. *Let C be a Gaussian copula of a normally distributed random vector $\mathbf{X}$ with positive defined covariance matrix. Then for every number $\alpha \in (0, 2]$ there exists an α-stable random vector $\mathbf{W}$ such that C is also the copula of $\mathbf{W}$.*

Proof. Because both addition and multiplication by positive numbers are strictly increasing transformations in $\mathbb{R}$, by virtue of Proposition 1 it can be supposed that all marginals $X_1, ..., X_d$ of the random vector $\mathbf{X}$ are standard normal random variables, $X_k \sim N(0, 1)$ for $k = 1, ..., d$. Based on the assumption, the covariance matrix Σ of $\mathbf{X}$ is positive defined, Lemma 3 implies the existence of an orthogonal $d \times d$ matrix $A = (a_{ij})$; $A^{-1} = A^T$, such that the normal random vector $\mathbf{Y} = A\mathbf{X}$ has independent marginals $Y_1, ..., Y_d$ and diagonal covariance matrix $A\Sigma A^T$ with diagonal elements consist of all eigenvalues of Σ.

Now α-stable random variables $S_1, ..., S_d$ are concerned. Lemma 1 ensures that, for each $k = 1, ..., d$, there exists a strictly increasing function $g_k : \mathbb{R} \to ran(S_k)$ such that the random variable $U_k = g_k \circ Y_k$ has the same α-stable distribution as S_k. Simultaneously, the independence of marginals $Y_1, ..., Y_d$ implies the independence of random variables $U_1, ..., U_d$. Thus, the random vector $\mathbf{U} = (U_1, ..., U_d)^T$ has independent α-stable marginals, it must be an α-stable random vector as the conclusion of Lemma 2. Let define a new random vector $\mathbf{W} = (W_1, ..., W_d)^T = A^{-1}\mathbf{U}$. Then it is clear that $\mathbf{W}$ is also an α-stable random vector as $\mathbf{U}$. We attempt to point out that $\mathbf{W}$ has the same copula C as $\mathbf{X}$. Firstly, the α-stability of marginals $W_1, ..., W_n$ of the random vector $\mathbf{W}$ together with Lemma 1 ensures that for each $k = 1, ..., d$, there exist strictly increasing function hk such that the random variable $Z_k = h_k \circ W_k$ has standard normal distribution as X_k. Consequently, due to Proposition 1, the random vector's $\mathbf{W}$ and $\mathbf{Z} = (Z_1, ..., Z_d)^T$ have the same copula, $C_{\mathbf{W}} = C_{\mathbf{Z}}$. Therefore, to complete the proof, it is sufficient to show that $C_{\mathbf{Z}} = C_{\mathbf{X}}$, which is equivalent to

$$c_{\mathbf{X}} = c_{\mathbf{Z}} \tag{10}$$

However, for $(u_1, ..., u_n) \in [0,1]^d$, it implies from (1) that

$$c_{\mathbf{X}}(u_1, ..., u_d) = \frac{f_{\mathbf{X}}(F_{X_1}^{-1}(u_1), ..., F_{X_d}^{-1}(u_d))}{f_{X_1}(F_{X_1}^{-1}(u_1))...f_{X_d}(F_{X_d}^{-1}(u_d))}$$

$$= \frac{f_{\mathbf{X}}(\Phi^{-1}(u_1), ..., \Phi^{-1}(u_d))}{\varphi(\Phi^{-1}(u_1))...\varphi(\Phi^{-1}(u_d))} \qquad (11)$$

and

$$c_{\mathbf{Z}}(u_1, ..., u_d) = \frac{f_{\mathbf{Z}}(F_{Z_1}^{-1}(u_1), ..., F_{Z_d}^{-1}(u_d))}{f_{Z_1}(F_{Z_1}^{-1}(u_1))...f_{Z_d}(F_{Z_d}^{-1}(u_d))}$$

$$= \frac{f_{\mathbf{Z}}(\Phi^{-1}(u_1), ..., \Phi^{-1}(u_d))}{\varphi(\Phi^{-1}(u_1))...\varphi(\Phi^{-1}(u_d))} \qquad (12)$$

From (11) and (12), it is evident that (10) is equivalent to

$$f_{\mathbf{X}}(\Phi^{-1}(u_1), ..., \Phi^{-1}(u_d)) = f_{\mathbf{Z}}(\Phi^{-1}(u_1), ..., \Phi^{-1}(u_d)) \qquad (13)$$

Denoting $\mathbf{g} := (g_1, ..., g_d)$, $\mathbf{h} := (h_1, ..., h_d)$ and

$$Q := A^{-1} \circ \mathbf{g} \circ A$$

we see that $\mathbf{Z} = \mathbf{h}(\mathbf{W})$ and $\mathbf{W} = Q(\mathbf{X})$. Then, due to $h_k = \Phi^{-1} \circ F_{W_k}$ and the independence of $Y_1, ..., Y_d$, setting $x_1 = F_{W_1}^{-1}(u_1), ..., x_d = F_{W_d}^{-1}(u_n)$, the right hand side of (13) equals to

$$\begin{aligned}
f_{\mathbf{Z}}(\Phi^{-1}(u_1), ..., \Phi^{-1}(u_d)) &= f_{\mathbf{h} \circ \mathbf{w}}(\Phi^{-1}(u_1), ..., \Phi^{-1}(u_d)) \\
&= f_{\mathbf{w}}(\mathbf{h}^{-1}(\Phi^{-1}(u_1), ..., \Phi^{-1}(u_d))) \\
&= f_{Q \circ \mathbf{x}}(h_1^{-1}(\Phi^{-1}(u_1)), ..., h_d^{-1}(\Phi^{-1}(u_d))) \\
&= f_{\mathbf{x}}(Q^{-1}(h_1^{-1}(\Phi^{-1}(u_1)), ..., h_d^{-1}(\Phi^{-1}(u_d)))) \\
&= f_{\mathbf{x}}(A^{-1} \circ \mathbf{g}^{-1} \circ A(h_1^{-1}(\Phi^{-1}(u_1)), ..., h_d^{-1}(\Phi^{-1}(u_d)))) \\
&= f_{\mathbf{x}}(A^{-1} \circ \mathbf{g}^{-1} \circ A(h_1^{-1}(F_{W_1}^{-1}(u_1)), ..., h_d^{-1}(F_{W_d}^{-1}(u_d)))) \\
&= f_{A^{-1} \circ \mathbf{Y}}(A^{-1} \circ \mathbf{g}^{-1} \circ A(x_1, ..., x_d)) \\
&= f_{\mathbf{Y}}(A \circ A^{-1} \circ \mathbf{g}^{-1} \circ A(x_1, ..., x_d)) \\
&= f_{Y_1}(g_1^{-1}(\sum_{j=1}^{d} a_{1j}x_j)) \times ... \times f_{Y_d}(g_d^{-1}(\sum_{j=1}^{d} a_{dj}x_j))
\end{aligned}$$

or

$$f_{\mathbf{Z}}(\Phi^{-1}(u_1), ..., \Phi^{-1}(u_d)) = f_{U_1}(\sum_{j=1}^{d} a_{1j}x_j) \times ... \times f_{U_d}(\sum_{j=1}^{d} a_{dj}x_j) \qquad (14)$$

Simultaneously, setting $y_1 = F_{U_1}^{-1}(u_1), ..., y_d = F_{U_d}^{-1}(u_d)$, the left hand side of (13) is equal to

$$
\begin{aligned}
f_{A^{-1} \circ \mathbf{Y}}(\Phi^{-1}(u_1), ..., \Phi^{-1}(u_d)) &= f_{\mathbf{Y}}(A(\Phi^{-1}(F_{U_1}(u_1)), ..., \Phi^{-1}(F_{U_d}(u_d)))) \\
&= f_{\mathbf{Y}}(A(g_1^{-1}(y_1), ..., g_d^{-1}(y_d))) \\
&= f_{Y_1}(\sum_{j=1}^{d} a_{1j} g_j^{-1}(y_j)) \times ... \times f_{Y_d}(\sum_{j=1}^{d} a_{dj} g_j^{-1}(y_j)) \\
&= \frac{1}{\sqrt{2\pi}} exp(-\frac{1}{2}[\sum_{j=1}^{d} a_{1j} g_j^{-1}(y_j)]^2) \times ... \\
&\quad ... \times \frac{1}{\sqrt{2\pi}} exp(-\frac{1}{2}[\sum_{j=1}^{d} a_{dj} g_j^{-1}(y_j)]^2) \\
&= (\frac{1}{\sqrt{2\pi}})^d exp(-\frac{1}{2} \sum_{k=1}^{d} [\sum_{j=1}^{d} a_{kj} g_j^{-1}(y_j)]^2) \\
&= (\frac{1}{\sqrt{2\pi}})^d exp(-\frac{1}{2} \sum_{k=1}^{d} \sum_{j=1}^{d} a_{kj} g_j^{-1}(y_j) \sum_{i=1}^{d} a_{ki} g_i^{-1}(y_i)) \\
&= (\frac{1}{\sqrt{2\pi}})^d exp(-\frac{1}{2} \sum_{k=1}^{d} \sum_{j=1}^{d} \sum_{i=1}^{d} a_{kj} a_{ki} g_j^{-1}(y_j) g_i^{-1}(y_i)) \\
&= (\frac{1}{\sqrt{2\pi}})^d exp(-\frac{1}{2} \sum_{j=1}^{d} \sum_{i=1}^{d} \delta_{ij} g_j^{-1}(x_j) g_i^{-1}(x_i)) \\
&= (\frac{1}{\sqrt{2\pi}})^d exp(-\frac{1}{2} \sum_{j=1}^{d} (g_j^{-1}(x_j))^2) \\
&= \varphi(g_1^{-1}(x_1)) \times ... \times \varphi(g_d^{-1}(x_d)) \\
&= f_{Y_1}(g_1^{-1}(y_1)) \times ... \times f_{Y_d}(g_d^{-1}(y_d))
\end{aligned}
$$

or

$$
f_{A^{-1} \circ \mathbf{Y}}(\Phi^{-1}(u_1), ..., \Phi^{-1}(u_d)) = f_{U_1}(y_1) \times ... \times f_{U_d}(y_d) \tag{15}
$$

where $\delta_{kk} = 1$ for $k = 1, ..., d, \delta_{ki} = 0$ for $k \neq i = 1, ..., d$. Comparing (14) to (15), with $y_1 = \sum_{j=1}^{d} a_{1j} x_j, ..., y_d = a_{dj} x_j$, we can conclude (3) is true, that means (10) holds, the proof completes. □

The next theorem presents Gaussian copula as a sufficient condition for a random vector with stable marginals to have stable distribution.

Theorem 3. *For given $\alpha \in (0, 2]$, let $\mathbf{X}$ be a Gaussian random vector with positive defined covariance matrix such that the matrix $A = (a_{ij})$ defined in the Lemma 3 satisfies $\det|(a_{ij})|^\alpha \neq 0$. Suppose that $\mathbf{W}^\star = (W_1^\star, ..., W_d^\star)^T$ is a random vector with α-stable marginals $W_1^\star, ..., W_d^\star$ such that the copula of $\mathbf{W}^\star$ and of $\mathbf{X}$ are equal, $C_{\mathbf{W}^\star} = C_{\mathbf{X}}$. Then $W^\star$ is an α-stable random vector.*

Proof. Let $\beta_1^\star, ..., \beta_d^\star \in [-1, 1]$, $\gamma_1^\star, ..., \gamma_d^\star$ and $\delta_1^\star, ..., \delta_d^\star \in \mathbb{R}$ be the stable parameters and $W_1^\star \sim S(\alpha, \beta_1^\star, \gamma_1^\star, \delta_1^\star), ..., W_d^\star \sim S(\alpha, \beta_d^\star, \gamma_d^\star, \delta_d^\star)$. Then, since α-stability of a random vector is unchanged after invertible linear transformation

$$(x_1, ..., x_d) \mapsto (a_1 x_1 + b_1, ..., a_d x_d + b_d)$$

for any $a_1 > 0, ..., a_d > 0$ and $(b_1, ..., b_d) \in \mathbb{R}^d$, without loss of generality, it can be supposed that $\gamma_1^\star = ... = \gamma_d^\star = 1$ and $\delta_1^\star = ... = \delta_d^\star = 0$

Let matrix $A = (a_{ij})$ and the random vector $Y = (Y_1, ..., Y_d)^T = AX$ be determined as in Lemma 3. We attempt to determine α-stable random variables $S_1 \sim S(\alpha, \beta_1, \gamma_1, 0), ..., S_d \sim S(\alpha, \beta_d, \gamma_d, 0)$ satisfying the equations

$$\begin{cases} 1 = (\gamma_1^\star)^\alpha = |a_{11}|^\alpha \gamma_1^\alpha + ... + |a_{d1}|^\alpha \gamma_d^\alpha \\ ... \\ 1 = (\gamma_1^\star)^\alpha = |a_{1d}|^\alpha \gamma_1^\alpha + ... + |a_{dd}|^\alpha \gamma_d^\alpha \end{cases} \tag{16}$$

and

$$\begin{cases} \beta_1^\star = \frac{\beta_1 |a_{11}|^\alpha \gamma_1^\alpha + ... + \beta_d |a_{d1}|^\alpha \gamma_d^\alpha}{|a_{11}|^\alpha \gamma_1^\alpha + ... + |a_{d1}|^\alpha \gamma_d^\alpha} = |a_{11}|^\alpha \gamma_1^\alpha \beta_1 + ... + |a_{d1}|^\alpha \gamma_d^\alpha \beta_d \\ ... \\ \beta_d^\star = \frac{\beta_1 |a_{1d}|^\alpha \gamma_1^\alpha + ... + \beta_d |a_{dd}|^\alpha \gamma_d^\alpha}{|a_{1d}|^\alpha \gamma_1^\alpha + ... + |a_{dd}|^\alpha \gamma_d^\alpha} = |a_{1d}|^\alpha \gamma_1^\alpha \beta_1 + ... + |a_{dd}|^\alpha \gamma_d^\alpha \beta_d \end{cases} \tag{17}$$

with unknowns $\gamma_1^\alpha, ..., \gamma_d^\alpha$ and $\gamma_1^\alpha \beta_1, ..., \gamma_d^\alpha \beta_d$.

From the assumption, $det|(a_{ij})|^\alpha \neq 0$, it is clear that the Eqs. (16) and (17) are solved and the α-stable random variables $S_1 \sim S(\alpha, \beta_1, \gamma_1, 0), ..., S_d \sim S(\alpha, \beta_d, \gamma_d, 0)$ are completely defined. By virtue of Corollary 2, for each $k = 1, ..., d$, there exists a strictly increasing function $g_k : \mathbb{R} \to rank(S_k)$ such that the random variable $U_k = g_k \circ Y_k$ has the same α-stable distribution as S_k. Simultaneously, the independence of marginals $Y_1, ..., Y_d$ implies the independence of α-stable random variables $U_1, ..., U_d$.

With $\mathbf{W} = (W_1, ..., W_d)^T = A^{-1}\mathbf{U}$, by the same argument of Theorem 2, it is certain that $\mathbf{W}$ is an α-stable random vector that has the same copula as $\mathbf{X}$; $C_{\mathbf{W}} = C_{\mathbf{X}} = C_{\mathbf{W}\star}$. On the other hand, the Eqs. (16) and (17) together with Proposition 1 imply the equality in distribution of all the marginals of $\mathbf{W}$ to correspondent marginals of $\mathbf{W}^\star$. In particular, $W_1 \sim S(\alpha, \beta_1^\star, \gamma_1^\star, \delta_1^\star), ..., W_d \sim S(\alpha, \beta_d^\star, \gamma_d^\star, \delta_d^\star)$. Consequently, $\mathbf{W}^\star \overset{D}{=} \mathbf{W}$ and $\mathbf{W}^\star$ is an α-stable random vector, the proof is fulfilled. $\qquad \square$

The above theorem suggests a procedure to check whether a data set can be fitted to any stable random vector or not, with details as follows:

Step 1. To estimate stable parameters of data marginals and to check if the all marginals have α-stable distributions with a common suitably chosen α.

Step 2. To estimate a Gaussian copula for the transformed data having normal distributed marginals, and to check if the original data are fitted to that Gaussian copula.

If the two steps are satisfied, it can be concluded the data vector has stable distribution with the stability index α.

Moreover, if a random vector is α-stable, we can easily consider that vector through by linear combination, in particular, calculation density and quantile values. Although, most of the density functions of stable random variables are not closed form, many computer software now support this calculation by numerical methods. However, calculations for joint density function of stable random vectors are still difficult. Nolan [12] represented an approach to determine density functions of stable random vectors belonging to a specific class of random vectors with elliptically contoured stable distributions, an α-stable radially symmetric (around 0) random vector $\mathbf{X}$.

However, it is clear that all marginals of an α-stable radially symmetric random vector are identically distributed and all marginals of an elliptically contoured stable random vector are symmetric (around respective location parameter). Those properties of data are quite rarely met in practice. In the following part, we propose a method of computing density functions of stable random vectors having Gaussian copulas. Whilst marginals of a stable random vector with Gaussian copula are not necessarily symmetric, this model of multivariate stable distributions may be more acceptable in the practical data analysis.

Let $\mathbf{G} \sim N(0, \Sigma)$ be a Gaussian random vector with covariance matrix $\Sigma = (\sigma_{ij})$ and $\sigma_{ii} = 1$ for all $i = 1, ..., d$. It is evident that Σ also is the correlation matrix of $\mathbf{G}$ and all marginals of this random vector are standard normal random variables. Then $\mathbf{G}$ has the density function

$$f_{\mathbf{G}}(\mathbf{x}) = (2\pi)^{-d/2}|\Sigma|^{-1/2}exp(-\frac{1}{2}\mathbf{x}\Sigma^{-1}\mathbf{x}^T)$$

for $\mathbf{x} = (x_1, ..., x_d) \in \mathbb{R}^d$, and its copula can be computed by

$$C_{\mathbf{G}}(\mathbf{u}) = (2\pi)^{-d/2}|\Sigma|^{-1/2} \int_{-\infty}^{\phi^{-1}(u_1)} ... \int_{-\infty}^{\phi^{-1}(u_d)} exp(-\frac{1}{2}\mathbf{x}\Sigma^{-1}\mathbf{x}^T)dx_d...dx_1$$

for $\mathbf{u} = (u_1, ..., u_d) \in [0, 1]^d$. Moreover from (1), its copula density is determined by

$$
\begin{aligned}
c_{\mathbf{G}}(\mathbf{t}) &= \frac{f_{\mathbf{G}}(\phi^{-1}(t_1), ..., \phi^{-1}(t_d))}{\varphi(\Phi^{-1}(t_1))...\varphi(\Phi^{-1}(t_d))} \\
&= \frac{exp(-\frac{1}{2}(\phi^{-1}(t_1), ..., \phi^{-1}(t_d))\Sigma^{-1}(\phi^{-1}(t_1), ..., \phi^{-1}(t_d))^T)}{(2\pi)^{d/2}|\Sigma|^{1/2}\varphi(\Phi^{-1}(t_1))...\varphi(\Phi^{-1}(t_d))}
\end{aligned}
\tag{18}
$$

for $\mathbf{t} = (t_1, ..., t_d) \in [0, 1]^d$, where φ, ϕ are pdf and cdf of standard normal random variable $N(0, 1)$, respectively:

$$\varphi(x) = \frac{1}{\sqrt{2\pi}}exp(\frac{-x^2}{2}), \phi(x) = \frac{1}{\sqrt{2\pi}} \int_{-\infty}^{x} exp(\frac{-t^2}{2})dt,$$

For fixed $\alpha \in (0, 2)$, let $S_1 \sim S(\alpha, \beta_1, \gamma_1, \delta_1), ..., S_d \sim S(\alpha, \beta_d, \gamma_d, \delta_d)$ be certain α-stable random variables with density functions $f_{S_1}, ..., f_{S_d}$ and cumulative distribution functions $F_{S_1}, ..., F_{S_d}$ respectively. Then formula (2) from Sklar's Theorem ensures the function

$$F_{\mathbf{Y}}(y_1, ..., y_d) = C_{\mathbf{G}}(F_{S_1}(y_1), ..., F_{S_d}(y_d))$$

for $y_1, ..., y_d \in \overline{\mathbb{R}} = [-\infty, +\infty]$, defines the cumulative distribution function of some random vector $\mathbf{Y}$ with copula $C_{\mathbf{G}}$ and marginals $S_1, ..., S_d$. In the case when the condition of Theorem 3 fulfilled for $A = \Sigma^{1/2}$, the random vector $\mathbf{Y}$ is truly an α-stable random vector, and by virtue of formula (1.4), its density function has the following

$$f_{\mathbf{Y}}(y_1, ..., y_d) = c_{\mathbf{G}}(F_{S_1}(y_1), ..., F_{S_d}(y_d)) \times f_{S_1}(y_1)...f_{S_d}(y_d)$$

$$= \frac{exp(-\frac{1}{2}(\phi^{-1}(F_{S_1}(y_1)), ..., \phi^{-1}(F_{S_d}(y_d)))\Sigma^{-1}(\phi^{-1}(F_{S_1}(y_1)), ..., \phi^{-1}(F_{S_d}(y_d)))^T)}{(2\pi)^{d/2}|\Sigma|^{1/2}\varphi(\Phi^{-1}(F_{S_1}(y_1)))...\varphi(\Phi^{-1}(F_{S_d}(y_d)))}$$

$$\times f_{S_1}(y_1)...f_{S_d}(y_d) \tag{19}$$

It is evident that (19) can be calculated by ordinary computing without symmetry assumption on the α-stable random variables $S_1, ..., S_d$. Therefore, the approach represented here is more flexible than the one of Nolan. The next section will show this application in stock market data of Vietnam.

4   Application for Vietnam Stock Market

In this section, we analyze data of stable distribution with Gaussian copula by using the results given in the previous section. The special structure of Gaussian copulas allows researchers to combine well - known computational tools for one-dimensional stable distributions and Gaussian copulas to compute density functions and quantile values of data which follow stable distributions with Gaussian copulas.

Dataset includes daily return of 4 stocks: BID (BIDVbank), VCB (Vietcombank), FLC (FLC Group), VNM (Vinamilk), with a sample from July 24, 2017 to October 14, 2019 to imply observations downloaded from Bao Viet Securities website. These are stocks on Ho Chi Minh Stock Exchange (abbreviated HOSE). Continuously compounded percentage returns are considered, i.e. daily returns are measured by log-differences of closing pricing multiplied by 100. Descriptive statistics together with Kolmogorov-Smirnov test for normal distribution of the univariate series are shown in Table 1 and the result for the univariate stable model estimation are presented in Table 2.

Table 1. Normal distribution test for HOSE daily returns data

Daily return	Size	Mean	SD	Skew	Kurtosis	p-value
BID	497	0.0015	0.03	0.42	3.74	0.00455
VCB	497	0.0021	0.02	0.34	3.85	0.00068
FLC	497	−0.0015	0.02	−1.18	13.55	0.00014
VNM	497	−0.0003	0.02	−2.47	43.59	0.00018

Table 2. K-S test for univariate stability of HOSE daily returns data

Daily return	α	$\bar{\alpha}$	β	γ	δ	p-value
BID	1.464,	1.416	0.008	0.0143	−0.0015	0.5046
VCB	1.446	1.416	0.164	0.0101	−0.0021	0.2884
FLC	1.256	1.416	0.185	0.0093	−0.0007	0.2233
VNM	1.498	1.416	0.072	0.0082	0.0002	0.1952

In Table 1, all p-values smaller than 5% confirm the significant divergence from normal distribution of the 4-dimensional vector. Simultaneously, the greater than 5% p-values of Kolmogorov-Smirnov tests in Table 2 are crucial arguments to conclude all the daily returns series of BID, VCB, FLC and VNM have univariate stable distributions with common stable index $\bar{\alpha} = 1.416$ (the average number of α's of those returns series).

After the above conclusion, we guess the 4-component vector of series of BID, VCB, FLC and VNM has multivariate stable distributions. To check this, we will use model of multivariate stable distributions with Gaussian copula. In the first step, the correlation matrices of daily returns (after normalizing by respective functions determined in Corollary 1) were calculated, with results given in Table 3 (showing only upper diagonal part of each correlation matrix). Then, the function named *gofCopula* in the copula package of R software was used to test the hypotheses of having Gaussian copula for 4-coordinates vector of daily returns, p-value is 0.2552. The result shows that the copula of daily returns vector is significantly Gaussian copula. Thus, the vector of 4 stocks BID, VCB, FLC and VNM has multivariate stable distributions with stable index $\bar{\alpha}$.

Table 3. Correlation matrices of yearly HOSE daily returns

Daily return	BID	VCB	FLC	VNM
BID	1	0.05001	0.00232	0.01380
VCB	-	1	0.04964	0.05349
FLC	-	-	1	0.00167
VNM	-	-	-	1

Consequently, (19) can be applied to determine the density functions of 4-coordinate vector of daily returns. In particular, the density functions are defined by the following explicit form:

$$f_{\mathbf{X}}(x_1, x_2, x_3, x_4) =$$

$$\frac{exp(-\frac{1}{2}(\Phi^{-1}(F_{X_1}(x_1)), ..., \Phi^{-1}(F_{X_4}(x_4))) \Sigma^{-1}(\Phi^{-1}(F_{X_1}(x_1)), ..., \Phi^{-1}(F_{X_4}(x_4)))^T}{(2\pi)^2 |\Sigma_i|^{1/2} \varphi(\Phi^{-1}(F_{X_1}(x_1)))...\varphi(\Phi^{-1}(F_{X_4}(x_4)))}$$

$$\times f_{X_1}(x_1)...f_{X_4}(x_4)$$

where $\mathbf{X} = (X_1, X_2, X_3, X_4)$ with 4-coordinate valued daily returns of stocks BID, VCB, FLC, and VNM respectively.

An investment portfolio is a set of financial assets owned by an investor that may include bonds, stocks, currencies, cash and cash equivalents, and commodities. Further, it refers to a group of investments that an investor uses in order to earn a profit while making sure that capital or assets are preserved, to achieve high returns and reduce the overall risk of the investment. To create a good investment portfolio, an investor or financial manager has to determine the objective of the portfolio. Markowitz model is a portfolio optimization model which was introduced in 1952 by Harry Markowitz [11]. In this model, Markowitz based on mean and variance of securities to choose securities give in investment portfolio. However, with multidimensional and stable distributed data, Markowitz model has complexity in computation and it is difficultly to decide. We use the feature which linear properties of α-stable random vector is univariate α-stable as a new tool which aims choosing optimal investment portfolio, that is calculating percentile values with given probabilities, comparing these values and choosing the maximum value.

In particular, we consider α-stable random vector $\mathbf{X} = (X_1, X_2, X_3, X_4)$ with 4-coordinate valued daily returns of stocks BID, VCB, FLC, and VNM respectively. According to [15], $\mathbf{X}$ is α-stable random vector, then every linear combination of coordinates $P = a_1 X_1 + a_2 X_2 + a_3 X_3 + a_4 X_4$ of vector $\mathbf{X}$ is also α-stable random variable. Thus, every investment portfolio $P = a_1 X_1 + a_2 X_2 + a_3 X_3 + a_4 X_4, a_1 + a_2 + a_3 + a_4 = 1$ has stable distribution with index $\bar{\alpha} = 1.416$. This result helps choosing investment portfolios more conveniently based on calculation percentile values of stable distribution. Table 4 shows percentile values with probabilities $1\%, 2\%, 3\%, 5\%$ in some investment portfolios.

Table 4. Percentile values of investment portfolios

Portfolio	(a_1, a_2, a_3, a_4)	1%	2%	3%	5%
1	$(\frac{1}{4}, \frac{1}{4}, \frac{1}{4}, \frac{1}{4})$	0.05258	0.03351	0.026087	0.01930
2	$(\frac{1}{2}, \frac{1}{6}, \frac{1}{6}, \frac{1}{6})$	0.08569	0.05400	0.04152	0.03001
3	$(\frac{1}{6}, \frac{1}{2}, \frac{1}{6}, \frac{1}{6})$	0.06852	0.04320	0.03325	0.02408
4	$(\frac{1}{6}, \frac{1}{6}, \frac{1}{2}, \frac{1}{6})$	0.06227	0.03918	0.03010	0.02172
5	$(\frac{1}{6}, \frac{1}{6}, \frac{1}{6}, \frac{1}{2})$	0.05180	0.03281	0.02540	0.01862

In Table 4, with the same probability, the 2nd investment portfolio has the maximum return value. Thus, the 2nd option is optimal for investment.

In short, the multivariate stable density functions can be directly used to compute the implied distribution of any portfolio of 4 assets BID, VCB, FLC and VNM. As the joint distribution of the vector of asset-returns is a multivariate stable distribution, the univariate distribution of returns of any portfolio of these assets is also stable. This approach can be used to solve many problems related to portfolio selection.

5 Conclusion and Future Works

Stable random vectors are increasingly being applied to real-world data, especially heavy-tailed data, while the calculation on these vectors is quite complicated. Besides the connection role, the Gaussian copula is also a tool to test vectors with stable distributions easily. However, this is just a rendezvous of the family of stable random vectors. Research can be extended to special classes on stable random vectors like sub-Gaussian stable distribution.

References

1. Embrechts, P., Lindskog, F., McNeil, A.: Modelling dependence with copulas and applications to risk management. In: Rachev, S. (ed.) Handbook of Heavy Tailed Distributions in Finance. Elsevier, Chapter 8, pp. 329–384 (2003)
2. Fama, E.: The behavior of stock prices. J. Bus. **38**, 34–105 (1965)
3. Fama, E., Roll, R.: Parameter estimates for symmetric stable distributions. J. Am. Stat. Assoc. **66**, 331–338 (1971)
4. Genest, C., Rémillard, B.: Validity of the parametric bootstrap for goodness-of-fit testing in semiparametric models. Ann. lInst. Henri Poincaré Prob. Statist. **44**(6), 1096–1127 (2008)
5. Kogon, S.M., Williams, D.B.: Characteristic Function Based Estimation of Stable Parameters, pp. 311–335. Birkhäuser, Boston (1998)
6. Kojadinovic, I., Yan, J.: Modeling multivariate distributions with continuous margins Using the copula R Package. J. Stat. Softw. **34**(9), 1–20 (2010)
7. Kunst, R.M.: Apparently stable increments in finance data: could ARCH effects be the cause?. J. Statist. Comput. Simulation **45**, 121–127 (1993)

8. Lamantia, F., Ortobelli, S., Rachev, S.: VaR, CVaR and time rules with elliptical and asymmetric stable distributed returns. Invest. Manag. Fin. Innov. **3**(4), 19–39 (2006)

9. McCulloch, J.H.: Financial applications of stable distributions. In: Maddala, G., Rao, C.: Handbook of Statistics, vol. 14, pp. 393–425. Elsevier Science Publishers, North-Holland (1996)

10. McCulloch, J.H.: Simple consistent estimators of stable distribution parameters: Comm. Statist. Simulation Comput. **15**, 1109–1136 (1986)

11. Markowitz, H.M.: Portfolio selection. J. Fin. **7** (1), 77–91 (1952)

12. Nolan, J.P.: Multivariate elliptically contoured stable distributions: theory and estimation. Comput. Statist. **28**(5), 2067–2089 (2013)

13. Nolan, J.P.: Maximum Likelihood Estimation and Diagnostics for Stable Distributions. pp. 379–400. Birkhäuser, Boston (2001)

14. Nolan, J.P., Panorska, A., McCulloch, J.H.: Estimation of stable spectral measures: Math. Comput. Model.**34**, 1113–1122 (2001)

15. Palmer, K.J., Ridout, M.S., Morgan, B.J.T.: Modelling cell generation times using the tempered stable distribution. J. R. Stat. Soc. Ser. C. Appl. Stat. **57**, 379–397 (2008)

16. Phuc, H.D., Truc Giang, V.T.: Gaussian copula of stable random vectors and application. Hacet. J. Math. Stat. **49**(2), 887–901 (2020)

17. Rachev, S.T.: Huang Xin: test for association of random variables in the domain of attraction of multivariate stable law. Probab. Math. Statist. **14**(1), 125–141 (1993)

18. Samuelson, P.: Efficient portfolio selection for Pareto - Lévy investments. J. Fin. Quant. Anal. **2**, 107–117 (1967)

19. Sklar, A.: Fonctions de répartition à n dimensions et leurs marges, vol. 8, pp. 229–231. Publications de l'Institut de Statistique de l'Université de Paris (1959)

20. Taqqu, M.S.: The modeling of Ethernet data and of signals that are heavy-tailed with infinite variance. Scand. J. Stat. **829**, 273–295 (2002)

21. Yu, J.: Empirical characteristic function estimation and its applications. Econ. Rev. **23**(2), 93–123 (2004)

22. Zolotarev, J.: Empirical characteristic function estimation and its applications. Econ. Rev. **23**(2), 93–123 (2004)

23. Adler, R.J., Feldman, R.E., Taqqu, M.S.: A Practical Guide to Heavy Tailed Data, Birkhäuser, Boston (1998)

24. Bui Quang, N.: On stable probability distributions and statistical application. Thesis, Academy of military science and technology, Ha Noi (2016)

25. Nolan, J.P.: Stable Distributions-Models for Heavy Tailed Data. Birkhauser, Boston (2016)

26. Hogg, R.V., McKean, J.W., Craig, A.T.: Introduction to Mathematical Statistics. Pearson Education, Inc., New York (2012)

27. Nelsen, R.P: An Introduction to Copulas. Springer, New York (2006). https://doi.org/10.1007/0-387-28678-0

28. Samorodnitsky, G., Taqqu, M.S.: Stable Non-Gaussian Random Processes, Chapman & Hall, New York (1994)

Numerical Results via High-Order Iterative Scheme to a Nonlinear Wave Equation with Source Containing Two Unknown Boundary Values

Pham Nguyen Nhat Khanh[1,2], Le Thi Mai Thanh[1,2,3],
Tran Trinh Manh Dung[1,2], and Nguyen Huu Nhan[3(✉)]

[1] Faculty of Mathematics and Computer Science, University of Science,
Ho Chi Minh City, 227 Nguyen Van Cu Street, District 5, Ho Chi Minh City, Vietnam
khanhm0715007@gstudent.ctu.edu.vn, ltmthanh@ntt.edu.vn
[2] Vietnam National University, Ho Chi Minh City, Vietnam
[3] Nguyen Tat Thanh University, 300A Nguyen Tat Thanh Street,
District 4, Ho Chi Minh City, Viet Nam
nhnhan@ntt.edu.vn

Abstract. In the present paper, we consider an initial-boundary value problem for nonlinear equation in which the source term contains two boundary values. First, the local existence and uniqueness of a weak solution to this problem is inferred directly from [12], in which a recurrent sequence via a N-order iterative scheme is established and then the N-order convergent rate of the sequence to the unique weak solution of the proposed model is also proved. As $N = 2$, the corresponding scheme called 2-order iterative scheme is considered, and a numerical algorithm to calculate approximate solutions via the 2-order iterative scheme is constructed by the finite-difference method. Finally, a numerical example is presented to evaluate the errors between the exact solution and the approximate solution. The numerical results show that the errors are decreasing as the fineness of meshes is increasing.

Keywords: Nonlinear wave equation · High-order iterative scheme · 2-order iterative scheme · Finite-difference scheme · Numerical results

Classification: 35L20 · 35L70 · 35Q72

1 Introduction

Our paper concerns with a Robin problem for a nonlinear equation as follows

$$u_{tt} - u_{xx} = f(x, t, u(x,t), \ u(0,t), \ u(1,t)), \ 0 < x < 1, \ 0 < t < T, \quad (1)$$

$$u_x(0,t) - h_0 u(0,t) = u_x(1,t) + h_1 u(1,t) = 0, \quad (2)$$

© ICST Institute for Computer Sciences, Social Informatics and Telecommunications Engineering 2021
Published by Springer Nature Switzerland AG 2021. All Rights Reserved
P. Cong Vinh and N. Huu Nhan (Eds.): ICTCC 2021, LNICST 408, pp. 209–224, 2021.
https://doi.org/10.1007/978-3-030-92942-8_17

$$u(x,0) = \tilde{u}_0(x), \quad u_t(x,0) = \tilde{u}_1(x), \tag{3}$$

where $h_0 \geq 0$, $h_1 \geq 0$ are constants, with $h_0 + h_1 > 0$ and f, $\tilde{u}_0$, $\tilde{u}_1$ are given functions.

For a approach of mathematical model, due to the presence of the unknown values $u(0,t)$ and $u(1,t)$ in the nonlinear terms, we take into account the fact that the problem (1)–(3) is related to nonlocal problems arised naturally in some applied sciences such as fluid mechanics, heat transfer theory and control theory, see [4,5]. The early studies of this type were considered by Agarwal [1], and later by Gupta [2] and Il'in and Moiseev [3], and the references cited therein.

We consider a recurrent sequence $\{u^{(m)}\}$ associated with the Eq. (1) (see [12]) as follows

$$u_0 \equiv 0,$$

$$\frac{\partial^2 u^{(m)}}{\partial t^2} - \Delta u^{(m)}$$

$$= \sum_{0 \leq i+j+s \leq N-1} D^{ijs} f[u^{(m-1)}] \left(u^{(m)} - u^{(m-1)}\right)^i \left(u^{(m)}(0,t) - u^{(m-1)}(0,t)\right)^j$$

$$\times \left(u^{(m)}(1,t) - u^{(m-1)}(1,t)\right)^s, \quad 0 < x < 1, \ 0 < t < T, \ m \geq 1, \tag{4}$$

where

$$D^{ijs} f[u^{(m-1)}](x,t) = \frac{1}{i!j!s!} D_3^i D_4^j D_5^s f\left(x,t,u^{(m-1)}(x,t), u^{(m-1)}(0,t), u^{(m-1)}(1,t)\right), \tag{5}$$

and $u^{(m)}$ satisfies (2) and (3). The above sequence is called high-order iterative scheme if it converges to the weak solution u of (1)–(3) and satisfies the estimation $\left\|u^{(m)} - u\right\|_X \leq C \left\|u^{(m-1)} - u\right\|_X^N$, for some $C > 0$, all integer numbers $N \geq 2$ and X is a certain Banach space. This definition is derived from the investigation of Newton-like methods in Banach spaces, see [13]. The high-order iterative scheme is also applied to some previous works, for example as in [9,10,12,16]. Especially, when $N = 2$, we get from (4) and (5) the following 2-order iterative scheme

$$\begin{cases} u_{tt}^{(m)} - u_{xx}^{(m)} = F^{(m)}(x,t), \ 0 < x < 1, \ 0 < t < T, \\ u_x^{(m)}(0,t) - h_0 u^{(m)}(0,t) = u_x^{(m)}(1,t) + h_1 u^{(m)}(1,t) = 0, \\ u^{(m)}(x,0) = \tilde{u}_0(x), \ u_t^{(m)}(x,0) = \tilde{u}_1(x), \end{cases} \tag{6}$$

where

$$F^{(m)}(x,t) = f[u^{(m-1)}](x,t) + D_3 f[u^{(m-1)}](x,t) \left(u^{(m)}(x,t) - u^{(m-1)}(x,t)\right)$$

$$+ D_4 f[u^{(m-1)}](x,t) \left(u^{(m)}(0,t) - u^{(m-1)}(0,t)\right) \tag{7}$$

$$+ D_5 f[u^{(m-1)}](x,t) \left(u^{(m)}(1,t) - u^{(m-1)}(1,t)\right).$$

In order to construct the numerical solutions of the 2-order iterative scheme (6)–(7), we shall use the spatial mesh $x_i = ih$, $h = 1/N$, $i = 0, 1, \cdots, N$, and Pinder's difference formulas [14] (pages 36, 43) to approximate the kth derivatives. Then the problem (6)–(7) will be transferred to a second-order system of ordinary equations (in time variable) of unknown functions $u_i^{(m)}(t) = u^{(m)}(x_i, t)$, $i = 0, 1, \cdots, N$, this method were also used in [7,8,11,15,17]. Next, by using the discretization with time mesh $t_j = j\Delta t$, $\Delta t = T/M$, $j = 0, 1, \cdots, M$, we will get an algorithm to determine the approximate solutions of $u^{(m)}$ to the scheme (6)–(7), which is given by the following difference equation (see the formula (38))

$$\bar{u}_{j+1}^{(m)} = \left[2E - (\Delta t)^2 A_j^{(m)}\right] u_j^{(m)} - u_{j-1}^{(m)} + (\Delta t)^2 \bar{\delta}_j^{(m)}, j = 1, \cdots, M - 1, \quad (8)$$

where $A_j^{(m)}$ and $\bar{\delta}_j^{(m)}$ are defined as in (37). Finally, we will show a numerical example to evaluate the errors between the approximate solution and the exact solution.

The outline of the paper is as follows. In Sect. 2, we introduce some notations and modified lemmas. In Sect. 3, by using the finite-difference method, we establish a numerical algorithm to determine the finite-difference approximate solutions of $u^{(m)}$ to the iterative scheme (6)–(7). Moreover, a numerical example is presented to evaluate the errors between the approximate solution $u^{(m)}$ and the exact solution u_{ex}. In Sect. 4, the conclusion is presented to describe the main results of the paper.

2   Preliminaries

In this section, we present some materials that we shall use in order to present our results.

Let $\Omega = (0, 1)$, $Q_T = (0, 1) \times (0, T)$ and $\langle \cdot, \cdot \rangle$ be the scalar product in L^2, i.e.,

$$\langle u, v \rangle = \int_0^1 u(x)v(x)dx,$$

and the corresponding norm $\|\cdot\|$, i.e., $\|u\|^2 = \langle u, u \rangle$. Also, we denote $\|\cdot\|_X$ is the norm in the Banach space X, and $L^p(0, T; X)$, $1 \le p \le \infty$ for the Banach space of real functions $u : (0, T) \to X$ measurable with the corresponding norm $\|\cdot\|_{L^p(0,T;X)}$ defined by

$$\|u\|_{L^p(0,T;X)} = \left(\int_0^T \|u(t)\|_X^p \, dt\right)^{1/p} < \infty \ for 1 \le p < \infty,$$

and

$$\|u\|_{L^\infty(0,T;X)} = \underset{0<t<T}{ess \sup} \|u(t)\|_X \quad for \ p = \infty.$$

Let $u(t)$, $u'(t) = u_t(t) = \dot{u}(t)$, $u''(t) = u_{tt}(t) = \ddot{u}(t)$, $u_x(t) = \nabla u(t)$, $u_{xx}(t) = \Delta u(t)$, denote $u(x,t)$, $\dfrac{\partial u}{\partial t}(x,t)$, $\dfrac{\partial^2 u}{\partial t^2}(x,t)$, $\dfrac{\partial u}{\partial x}(x,t)$, $\dfrac{\partial^2 u}{\partial x^2}(x,t)$, respectively.

For $f \in C^k([0,1] \times \mathbb{R}_+ \times \mathbb{R}^3)$, $f = f(x,t,y_1,y_2,y_3)$, we put $D_1 f = \dfrac{\partial f}{\partial x}$, $D_2 f = \dfrac{\partial f}{\partial t}$, $D_{2+i} f = \dfrac{\partial f}{\partial y_i}$, $i = 1,2,3$ and $D^\alpha f = D_1^{\alpha_1} ... D_5^{\alpha_5} f$; $\alpha = (\alpha_1,...,\alpha_5) \in \mathbb{Z}_+^5$, $|\alpha| = \alpha_1 + ... + \alpha_5 = k$, $D^{(0,...,0)} f = D^{(0)} f = f$.

We consider the symmetric bilinear form $a(\cdot,\cdot)$ defined by

$$a(u,v) = \int_0^1 u_x(x) v_x(x) dx + h_0 u(0) v(0) + h_1 u(1) v(1), \text{ with } u, v \in H^1, \quad (9)$$

and the corresponding norm $\|v\|_a = \sqrt{a(v,v)}$.

Then, two norms $\|v\|_{H^1}$, and $\|v\|_a$ are equivalent norms on H^1. Furthermore, the following inequalities are valid, see [12] (Lemma 2.1 and 2.2)

$$
\begin{aligned}
&\text{(i)} \quad \|v\|_{C^0(\bar{\Omega})} \leq \sqrt{2} \|v\|_{H^1}, \\
&\text{(ii)} \quad \sqrt{a_0} \|v\|_{H^1} \leq \|v\|_a \leq \sqrt{a_1} \|v\|_{H^1}, \\
&\text{(iii)} \quad |a(u,v)| \leq a_1 \|u\|_{H^1} \|v\|_{H^1}, \text{ for all } u, v \in H^1, \\
&\text{(4i)} \quad a(v,v) \geq a_0 \|v\|_{H^1}^2, \text{ for all } v \in H^1,
\end{aligned}
\quad (10)
$$

where $a_0 = \dfrac{1}{3} \min\{1, \max\{h_0, h_1\}\}$ and $a_1 = 1 + 2(h_0 + h_1)$.

Definition 1. *A function $u = u(x,t)$ is a weak solution of the problem (1)–(3) if*

$$u \in L^\infty(0,T;H^2), \quad u' \in L^\infty(0,T;H^1), \quad u'' \in L^\infty(0,T;L^2),$$

and u satisfies the following variational equation

$$\langle u''(t), w \rangle + a(u(t), w) = \langle f[u](t), w \rangle, \forall w \in H^1, \text{ and a.e., } t \in (0,T), \quad (11)$$

where $f[u](x,t) = f(x,t,u(x,t), u(0,t), u(1,t))$, together with the initial conditions

$$u(0) = \tilde{u}_0, \quad u'(0) = \tilde{u}_1. \quad (12)$$

For our study of problem (1)–(3) we will need the following assumptions

(H_1) $(\tilde{u}_0, \tilde{u}_1) \in H^2 \times H^1$;

(H_2) $f \in C^0([0,1] \times \mathbb{R}_+ \times \mathbb{R}^3)$ such that

(i) $D_3^i D_4^j D_5^s f \in C^0([0,1] \times \mathbb{R}_+ \times \mathbb{R}^3)$, with $0 \leq i + j + s \leq 2$,

(ii) $D_1 D_3^i D_4^j D_5^s f$, $D_3^{i+1} D_4^j D_5^s f \in C^0([0,1] \times \mathbb{R}_+ \times \mathbb{R}^3)$, with $i + j + s = 1$.

Let fixed positive constant T^* and positive constant M. For every $T \in (0, T^*]$, we put

$$W_T = \{v \in L^\infty(0,T;H^2) : v_t \in L^\infty(0,T;H^1), \ v_{tt} \in L^2(Q_T)\}. \quad (13)$$

Note that W_T is a Banach space with respect to the norm

$$\|v\|_{W_T} = \max\{\|v\|_{L^\infty(0,T;H^2)}, \|v_t\|_{L^\infty(0,T;H^1)}, \|v_{tt}\|_{L^2(Q_T)}\} \tag{14}$$

(see Lions [6]). We also put

$$\begin{cases} W(M,T) = \{v \in W_T : \|v\|_{W_T} \leq M\}, \\ W_1(M,T) = \{v \in W(M,T) : v_{tt} \in L^\infty(0,T;L^2)\}. \end{cases} \tag{15}$$

To prove the existence and the uniqueness of solution to the problem (1)–(3), we cite the results given in [12] where the authors used the high-order iterative method to construct a recurrence sequence converging, at N-order rate, to the weak solutions of the problem (1)–(3). Therefore, in case of $N = 2$, the existence and the convergence of the recurrence $\{u^{(m)}\}$ defined by (6)–(7) are claimed the following theorems.

Theorem 1. *Let (H_1)–(H_2) hold. Then there are two positive constants M and T depending on $\tilde{u}_0$, $\tilde{u}_1$, f, h_0, h_1, such that, for $u_0 \equiv 0$, the recurrent sequence $\{u^{(m)}\}$ defined by (6)–(7) exists and contains in $W_1(M,T)$.*

Using Theorem 1 and some arguments of weak convergence, the following theorem is also confirmed.

Theorem 2. *Let (H_1) – (H_2) hold. Then, there are two positive constants M and T such that the problem (1)–(3) has a unique weak solution $u \in W_1(M,T)$ and the recurrent sequence $\{u^{(m)}\}$ defined by (6)–(7) converges, at 2-order rate, to u strongly in the space $W_1(T)$ in sense*

$$\left\|u^{(m)} - u\right\|_{W_1(T)} \leq C \left\|u^{(m-1)} - u\right\|_{W_1(T)}^2, \tag{16}$$

for all $m \geq 1$, where C is a suitable constant. On the other hand, the following estimate is fulfilled

$$\left\|u^{(m)} - u\right\|_{W_1(T)} \leq C_T(k_T)^{2^m}, \text{ for all } m \in \mathbb{N}, \tag{17}$$

where $C_T > 0$ and $0 < k_T < 1$ are constants depending only on T.

3   Numerical Results

Consider the following problem

$$\begin{cases} u_{tt} - u_{xx} = f(x,t,u(x,t),u(0,t),u(1,t)), \ 0 < x < 1, \ 0 < t < T, \\ u_x(0,t) - h_0 u(0,t) = u_x(1,t) + h_1 u(1,t) = 0, \\ u(x,0) = \tilde{u}_0(x), \ u_t(x,0) = \tilde{u}_1(x), \end{cases} \tag{18}$$

where

$$
\begin{cases}
f(x,t,u(x,t),u(0,t),u(1,t)) = \dfrac{1}{16}|u|u + \dfrac{25}{16}u^2(0,t) + \dfrac{125}{27}u^3(1,t) + G(x,t), \\[2mm]
h_0 = 1, h_1 = 2, \\[2mm]
G(x,t) = -e^{-\frac{3t}{2}} + \dfrac{1}{4}\left(-x^2 + \dfrac{4}{5}x + \dfrac{44}{5}\right)e^{\frac{-t}{2}} \\[4mm]
\qquad\qquad -\dfrac{1}{16}\left(x^4 - \dfrac{8}{5}x^3 - \dfrac{24}{25}x^2 + \dfrac{32}{25}x + \dfrac{416}{25}\right)e^{-t} \\[4mm]
\tilde{u}_0(x) = -x^2 + \dfrac{4}{5}x + \dfrac{4}{5}, \quad \tilde{u}_1(x) = \dfrac{1}{2}x^2 - \dfrac{2}{5}x - \dfrac{2}{5}
\end{cases}
$$

$$(19)$$

Then $u_{ex}(x,t) = \left(-x^2 + \dfrac{4}{5}x + \dfrac{4}{5}\right)e^{\frac{-1}{2}t}$ is the exact solution of (18) corresponding to the constants $h_0 = 1$, $h_1 = 2$ and the given functions $f(x,t,u(x,t),u(0,t),u(1,t))$, $\tilde{u}_0$, $\tilde{u}_1$ as in (19).

In this section, we first construct a difference scheme to approximate the solution u of (18) via approximating $u^{(m)}$ in the 2-order iterative scheme as follows

$$
\begin{cases}
u_{tt}^{(m)} - u_{xx}^{(m)} = F^{(m)}(x,t), \ 0 < x < 1, \ 0 < t < T, \\[2mm]
u_x^{(m)}(0,t) - h_0 u^{(m)}(0,t) = u_x^{(m)}(1,t) + h_1 u^{(m)}(1,t) = 0, \\[2mm]
u^{(m)}(x,0) = \tilde{u}_0(x), \ u_t^{(m)}(x,0) = \tilde{u}_1(x),
\end{cases}
$$

$$(20)$$

where

$$
\begin{aligned}
F^{(m)}(x,t) &= f[u^{(m-1)}](x,t) + D_3 f[u^{(m-1)}](x,t)\left(u^{(m)}(x,t) - u^{(m-1)}(x,t)\right) \\
&\quad + D_4 f[u^{(m-1)}](x,t)\left(u^{(m)}(0,t) - u^{(m-1)}(0,t)\right) \\
&\quad + D_5 f[u^{(m-1)}](x,t)\left(u^{(m)}(1,t) - u^{(m-1)}(1,t)\right) \\
&= \alpha^{(m)}(x,t)u^{(m)}(x,t) + \beta^{(m)}(x,t)u^{(m)}(0,t) \\
&\quad + \gamma^{(m)}(x,t)u^{(m)}(1,t) + \delta^{(m)}(x,t),
\end{aligned}
$$

$$(21)$$

and

$$
\begin{aligned}
\alpha^{(m)}(x,t) &= D_3 f[u^{(m-1)}](x,t), \\
\beta^{(m)}(x,t) &= D_4 f[u^{(m-1)}](x,t), \\
\gamma^{(m)}(x,t) &= D_5 f[u^{(m-1)}](x,t), \\
\delta^{(m)}(x,t) &= f[u^{(m-1)}](x,t) - \alpha^{(m)}(x,t)u^{(m-1)}(x,t) \\
&\quad - \beta^{(m)}(x,t)u^{(m-1)}(0,t) - \gamma^{(m)}(x,t)u^{(m-1)}(1,t),
\end{aligned}
$$

$$(22)$$

with the notations

$$f[u](x,t) = f(x,t,u(x,t),u(0,t),u(1,t)),$$
$$D_i f[u](x,t) = D_i f(x,t,u(x,t),u(0,t),u(1,t)), \; i = 3,4,5. \qquad (23)$$

Replacing (21) and (22) into (20), we get

$$
\begin{cases}
u_{tt}^{(m)} - u_{xx}^{(m)} - \alpha^{(m)}(x,t)u^{(m)} - \beta^{(m)}(x,t)u^{(m)}(0,t) - \gamma^{(m)}(x,t)u^{(m)}(1,t) \\
\qquad\qquad = \delta^{(m)}(x,t), \; 0 < x < 1, \; 0 < t < T, \\
u_x^{(m)}(0,t) - h_0 u^{(m)}(0,t) = u_x^{(m)}(1,t) + h_1 u^{(m)}(1,t) = 0, \\
u^{(m)}(x,0) = \tilde{u}_0(x), \; u_t^{(m)}(x,0) = \tilde{u}_1(x).
\end{cases}
\qquad (24)
$$

Putting

$$u_i^{(m)}(t) = u^{(m)}(x_i,t), \; x_i = ih, \; i = 0,1,\cdots,N, \; h = \frac{1}{N}. \qquad (25)$$

Rewriting (24) at the node $x = x_i$:

$$
\begin{cases}
\ddot{u}_i^{(m)}(t) - \Delta u_i^{(m)}(t) - \alpha_i^{(m)}(t)u_i^{(m)}(t) - \beta_i^{(m)}(t)u_0^{(m)}(t) - \gamma_i^{(m)}(t)u_N^{(m)}(t) \\
\qquad\qquad = \delta_i^{(m)}(t), \; 0 \le i \le N, \; 0 \le t \le T, \\
u_x^{(m)}(0,t) - h_0 u_0^{(m)}(t) = u_x^{(m)}(1,t) + h_1 u_N^{(m)}(t) = 0, \\
u_i^{(m)}(0) = \tilde{u}_0(x_i) = \tilde{u}_{0i}, \; \dot{u}_i^{(m)}(0) = \tilde{u}_1(x_i) = \tilde{u}_{1i}, \; 0 \le i \le N,
\end{cases}
\qquad (26)
$$

in which

$$
\begin{aligned}
\alpha_i^{(m)}(t) &= \alpha^{(m)}(x_i,t) = D_3 f[u^{(m-1)}](x_i,t) \\
&= D_3 f(x_i,t,u_i^{(m-1)}(t),u_0^{(m-1)}(t),u_N^{(m-1)}) \qquad (27) \\
\beta_i^{(m)}(t) &= \beta^{(m)}(x_i,t) = D_4 f[u^{(m-1)}](x_i,t) \\
&= D_4 f(x_i,t,u_i^{(m-1)}(t),u_0^{(m-1)}(t),u_N^{(m-1)}(t)), \\
\gamma_i^{(m)}(t) &= \gamma^{(m)}(x_i,t) = D_5 f[u^{(m-1)}](x_i,t) \\
&= D_5 f(x_i,t,u_i^{(m-1)}(t),u_0^{(m-1)}(t),u_N^{(m-1)}(t)), \\
\delta_i^{(m)}(t) &= \delta^{(m)}(x_i,t) \\
&= f(x_i,t,u_i^{(m-1)}(t),u_0^{(m-1)}(t),u_N^{(m-1)}(t)) - \alpha_i^{(m)}(t)u_i^{(m-1)}(t) \\
&\quad - \beta_i^{(m)}(t)u_0^{(m-1)}(t) - \gamma_i^{(m)}(t)u_N^{(m-1)}(t).
\end{aligned}
$$

Replacing the derivatives in spatial variable x of (26) by the following approximations

$$u_x^{(m)}(x_i, t) \simeq \frac{u_i^{(m)}(t) - u_{i-1}^{(m)}(t)}{h}, \tag{28}$$

$$u_x^{(m)}(0, t) \simeq \frac{u_1^{(m)}(t) - u_0^{(m)}(t)}{h},$$

$$u_x^{(m)}(1, t) \simeq \frac{u_N^{(m)}(t) - u_{N-1}^{(m)}(t)}{h},$$

$$u_{xx}^{(m)}(x_i, t) \simeq \frac{u_{i-1}^{(m)}(t) - 2u_i^{(m)}(t) + u_{i+1}^{(m)}(t)}{h^2},$$

then we obtain

$$\ddot{u}_i^{(m)}(t) - \frac{u_{i-1}^{(m)}(t) - 2u_i^{(m)}(t) + u_{i+1}^{(m)}(t)}{h^2} - \alpha_i^{(m)}(t)u_i^{(m)}(t) - \beta_i^{(m)}(t)u_0^{(m)}(t)$$
$$-\gamma_i^{(m)}(t)u_N^{(m)}(t) = \delta_i^{(m)}(t), \ 1 \le i \le N-1,$$

$$\frac{u_1^{(m)}(t) - u_0^{(m)}(t)}{h} - h_0 u_0^{(m)}(t) = \frac{u_N^{(m)}(t) - u_{N-1}^{(m)}(t)}{h} + h_1 u_N^{(m)}(t) = 0,$$

$$u_0^{(m)}(t) = \frac{u_1^{(m)}(t)}{1 + hh_0}, \ u_N^{(m)}(t) = \frac{u_{N-1}^{(m)}(t)}{1 + hh_1},$$

$$u_i^{(m)}(0) = \tilde{u}_0(x_i) = \tilde{u}_{0i}, \ \dot{u}_i^{(m)}(0) = \tilde{u}_1(x_i) = \tilde{u}_{1i}, \ 0 \le i \le N, \tag{29}$$

Using the boundary conditions $(29)_2$ with $u_0^{(m)}(t) = \dfrac{u_1^{(m)}(t)}{1 + hh_0}$,

$u_N^{(m)}(t) = \dfrac{u_{N-1}^{(m)}(t)}{1 + hh_1}$, and after eliminating the unknown functions $u_0(t)$ and $u_N(t)$ in the equations $i = 1, 2, N-2, N-1$, then the system (29) is rewritten as follows

$$\begin{cases} \ddot{u}_1^{(m)}(t) + \hat{a}_1^{(m)}(t)u_1^{(m)}(t) - \dfrac{1}{h^2}u_2^{(m)}(t) + \bar{a}_1^{(m)}(t)u_{N-1}^{(m)}(t) = \delta_1^{(m)}(t), \\[2mm] \ddot{u}_2^{(m)}(t) + \hat{a}_2^{(m)}(t)u_1^{(m)}(t) + b_2^{(m)}(t)u_2^{(m)}(t) - \dfrac{1}{h^2}u_3^{(m)}(t) \\[2mm] \qquad\qquad + \bar{a}_2^{(m)}(t)u_{N-1}^{(m)}(t) = \delta_2^{(m)}(t), \\[2mm] \qquad \vdots \\[2mm] \ddot{u}_i^{(m)}(t) + \hat{a}_i^{(m)}(t)u_1^{(m)}(t) - \dfrac{1}{h^2}u_{i-1}^{(m)}(t) + b_i^{(m)}(t)u_i^{(m)}(t) - \dfrac{1}{h^2}u_{i+1}^{(m)}(t) \\[2mm] \qquad\qquad + \bar{a}_i^{(m)}(t)u_{N-1}^{(m)}(t) = \delta_i^{(m)}(t), \ 3 \le i \le N-3, \\[2mm] \qquad \vdots \\[2mm] \ddot{u}_{N-2}^{(m)}(t) + \hat{a}_{N-2}^{(m)}(t)u_1^{(m)}(t) - \dfrac{1}{h^2}u_{N-3}^{(m)}(t) + b_{N-2}^{(m)}(t)u_{N-2}^{(m)}(t) \\[2mm] \qquad\qquad + \bar{a}_{N-2}^{(m)}(t)u_{N-1}^{(m)}(t) = \delta_{N-2}^{(m)}(t), \\[2mm] \ddot{u}_{N-1}^{(m)}(t) + \hat{a}_{N-1}^{(m)}(t)u_1^{(m)}(t) - \dfrac{1}{h^2}u_{N-2}^{(m)}(t) + \bar{a}_{N-1}^{(m)}(t)u_{N-1}^{(m)}(t) = \delta_{N-1}^{(m)}(t), \end{cases} \tag{30}$$

where

$$
\hat{a}_i^{(m)}(t) = \begin{cases} -\dfrac{\beta_1^{(m)}(t)}{1+hh_0} - \alpha_1^{(m)}(t) - \dfrac{1}{h^2(1+hh_0)} + \dfrac{2}{h^2}, & i = 1, \\[3mm] -\dfrac{\beta_2^{(m)}(t)}{1+hh_0} - \dfrac{1}{h^2}, & i = 2, \\[3mm] -\dfrac{\beta_i^{(m)}(t)}{1+hh_0}, & 3 \le i \le N-1, \end{cases}
$$

$$
\bar{a}_i^{(m)}(t) = \begin{cases} -\dfrac{\gamma_i^{(m)}(t)}{1+hh_1}, & 1 \le i \le N-3, \quad (31) \\[3mm] -\dfrac{\gamma_{N-2}^{(m)}(t)}{1+hh_1} - \dfrac{1}{h^2}, & i = N-2, \\[3mm] -\dfrac{\gamma_{N-1}^{(m)}(t)}{1+hh_1} - \alpha_{N-1}^{(m)}(t) - \dfrac{1}{h^2(1+hh_1)} + \dfrac{2}{h^2}, & i = N-1, \end{cases}
$$

$$
b_i^{(m)}(t) = \frac{2}{h^2} - \alpha_i^{(m)}(t), \quad 2 \le i \le N-2.
$$

We rewrite (30) into a vector equation as follows

$$
\begin{cases} \ddot{X}^{(m)}(t) + A^{(m)}(t)X^{(m)}(t) = \vec{\delta}^{(m)}(t), \\ X^{(m)}(0) = X_0, \ \dot{X}^{(m)}(0) = X_1, \end{cases} \tag{32}
$$

where

$$
\begin{cases} X^{(m)}(t) = \left(u_1^{(m)}(t), \cdots, u_{N-1}^{(m)}(t) \right)^T, \\[2mm] \vec{\delta}^{(m)}(t) = \left(\delta_1^{(m)}(t), \cdots, \delta_{N-1}^{(m)}(t) \right)^T, \\[2mm] X_0 = \left(\tilde{u}_0(x_1), \cdots, \tilde{u}_0(x_{N-1}) \right)^T, \\[2mm] X_1 = \left(\tilde{u}_0(x_1), \cdots, \tilde{u}_0(x_{N-1}) \right)^T, \end{cases} \tag{33}
$$

and $A^{(m)}(t) \in \mathfrak{M}_{N-1}$ ($\mathfrak{M}_{N-1}$ is the set of real $N-1$-order matrices) is defined by

$$A^{(m)}(t) = \begin{bmatrix} \hat{a}_1^{(m)}(t) & \frac{-1}{h^2} & 0 & \cdots & \cdots & \cdots & \cdots & \cdots & 0 & \bar{a}_1^{(m)}(t) \\ \hat{a}_2^{(m)}(t) & b_2^{(m)}(t) & \frac{-1}{h^2} & 0 & \cdots & \cdots & \cdots & \cdots & 0 & \bar{a}_2^{(m)}(t) \\ \hat{a}_3^{(m)}(t) & \frac{-1}{h^2} & b_3^{(m)}(t) & \frac{-1}{h^2} & 0 & \cdots & \cdots & \cdots & \vdots & \vdots \\ \vdots & 0 & \ddots & \ddots & \ddots & \ddots & \cdots & \cdots & \vdots & \vdots \\ \hat{a}_i^{(m)}(t) & 0 & 0 & \frac{-1}{h^2} & b_i^{(m)}(t) & \frac{-1}{h^2} & 0 & \cdots & \vdots & \bar{a}_i^{(m)}(t) \\ \vdots & \vdots & \vdots & \ddots & \ddots & \ddots & \ddots & \ddots & \vdots & \vdots \\ \vdots & \vdots & \vdots & \cdots & \ddots & \ddots & \ddots & \ddots & 0 & \vdots \\ \hat{a}_{N-3}^{(m)}(t) & \vdots & \vdots & \cdots & \cdots & 0 & \frac{-1}{h^2} & b_{N-3}^{(m)}(t) & \frac{-1}{h^2} & \bar{a}_{N-3}^{(m)}(t) \\ \hat{a}_{N-2}^{(m)}(t) & \vdots & \vdots & \cdots & \cdots & \cdots & 0 & \frac{-1}{h^2} & b_{N-2}^{(m)}(t) & \bar{a}_{N-2}^{(m)}(t) \\ \hat{a}_{N-1}^{(m)}(t) & 0 & 0 & \cdots & \cdots & \cdots & \cdots & 0 & \frac{-1}{h^2} & \bar{a}_{N-1}^{(m)}(t) \end{bmatrix} \quad (34)$$

Approximating the derivatives $\dot{X}^{(m)}(t_j)$, $\ddot{X}^{(m)}(t_j)$ by the differences in time variable and the following partition

$$\dot{X}^{(m)}(t_j) \simeq \frac{X_{j+1}^{(m)} - X_j^{(m)}}{\Delta t}, \quad (35)$$

$$\ddot{X}^{(m)}(t_j) \simeq \frac{X_{j-1}^{(m)} - 2X_j^{(m)} + X_{j+1}^{(m)}}{(\Delta t)^2},$$

$$X_j^{(m)} = X^{(m)}(t_j) = \left(u_1^{(m)}(t_j), \cdots, u_{N-1}^{(m)}(t_j) \right)^T,$$

with $t_j = j\Delta t$, $j = 0, \cdots, M$, $\Delta t = \dfrac{T}{M}$,

$$X_1 = \dot{X}^{(m)}(0) \simeq \frac{X_1^{(m)} - X_0^{(m)}}{\Delta t},$$

then the Eq. (32) was rewritten as follows

$$\begin{cases} \dfrac{X_{j-1}^{(m)} - 2X_j^{(m)} + X_{j+1}^{(m)}}{(\Delta t)^2} + A_j^{(m)} X_j^{(m)} = \vec{\delta}_j^{(m)}, \\ X_0^{(m)} = X_0, \\ X_1^{(m)} = X_0^{(m)} + \Delta t X_1, \end{cases} \quad (36)$$

where

$$\vec{\delta}_j^{(m)} = \vec{\delta}^{(m)}(t_j) = \left(\delta_1^{(m)}(t_j), \cdots, \delta_{N-1}^{(m)}(t_j) \right)^T, \quad (37)$$

$$A_j^{(m)} = A^{(m)}(t_j).$$

Hence, the Eq. (36) can be rewritten as below

$$
\begin{cases}
X_{j+1}^{(m)} = \left[2E - (\Delta t)^2 A_j^{(m)} \right] X_j^{(m)} - X_{j-1}^{(m)} + (\Delta t)^2 \vec{\delta}_j^{(m)}, \; j = 1, \cdots, M-1, \\
X_0^{(m)} = X_0, \\
X_1^{(m)} = X_0 + \Delta t X_1,
\end{cases}
\tag{38}
$$

in which

$$
\begin{cases}
X^{(m)}(t) = \left(u_1^{(m)}(t), \cdots, u_{N-1}^{(m)}(t) \right)^T = \left(u^{(m)}(x_1, t), \cdots, u^{(m)}(x_{N-1}, t) \right)^T, \\
X_j^{(m)} = X^{(m)}(t_j) = \left(u^{(m)}(x_1, t_j), \cdots, u^{(m)}(x_{N-1}, t_j) \right)^T, \\
X_0 = \left(\tilde{u}_0(x_1), \cdots, \tilde{u}_0(x_{N-1}) \right)^T, \\
X_1 = \left(\tilde{u}_1(x_1), \cdots, \tilde{u}_1(x_{N-1}) \right)^T, \\
\vec{\delta}^{(m)}(t) = \left(\delta_1^{(m)}(t), \cdots, \delta_{N-1}^{(m)}(t) \right)^T, \\
\vec{\delta}_j^{(m)} = \vec{\delta}^{(m)}(t_j) = \left(\delta_1^{(m)}(t_j), \cdots, \delta_{N-1}^{(m)}(t_j) \right)^T, \\
\delta_i^{(m)}(t_j) = \delta^{(m)}(x_i, t_j) \\
\qquad = f(x_i, t_j, u_i^{(m-1)}(t_j), u_0^{(m-1)}(t_j), u_N^{(m-1)}(t_j)) - \alpha_i^{(m)}(t_j) u_i^{(m-1)}(t_j) \\
\qquad - \beta_i^{(m)}(t_j) u_0^{(m-1)}(t_j) - \gamma_i^{(m)}(t_j) u_N^{(m-1)}(t_j), \\
u_0^{(m-1)}(t) = \dfrac{u_1^{(m-1)}(t)}{1 + h h_0}, \; u_N^{(m)}(t) = \dfrac{u_{N-1}^{(m-1)}(t)}{1 + h h_1},
\end{cases}
\tag{39}
$$

We describe the scheme (38) as follows.

A. For fixed constants M, N. With $m = 0$, we give the first vector

$$
X_j^{(0)} = \left(u_1^{(0)}(t_j), \cdots, u_{N-1}^{(0)}(t_j) \right)^T = \left(u^{(0)}(x_1, t_j), \cdots, u^{(0)}(x_{N-1}, t_j) \right)^T \equiv 0, \; j = 1, \cdots, M.
$$

B. At the $(m-1)^{\text{th}}$ iterative step, suppose that we know

$$
X_j^{(m-1)} = X^{(m-1)}(t_j) = \left(u_1^{(m-1)}(t_j), \cdots, u_{N-1}^{(m-1)}(t_j) \right)^T, \; j = 1, \cdots, M.
$$

C. Then, we compute consecutively the vectors $X_j^{(m)} = \left(u_1^{(m)}(t_j), \cdots, \right.$ $\left. u_{N-1}^{(m)}(t_j) \right)^T, \; j = 1, \cdots, M$ by the steps as below

C1. The vector $X_1^{(m)}$ is defined by $X_1^{(m)} = \left(u_1^{(m)}(t_1), \cdots, u_{N-1}^{(m)}(t_1) \right)^T =$ $X_0 + \Delta t X_1.$

C2. The vector $X_2^{(m)}$ is defined by $X_2^{(m)} = \left(u_1^{(m)}(t_2), \cdots, u_{N-1}^{(m)}(t_2) \right)^T$.

(i) For the vectors $X_0 = \left(\tilde{u}_0(x_1), \cdots, \tilde{u}_0(x_{N-1}) \right)^T, X_1 = \left(\tilde{u}_1(x_1), \cdots, \tilde{u}_1(x_{N-1}) \right)^T$, we compute the values $\hat{a}_i^{(m)}(t_1), (2 \le i \le N - 2), \bar{a}_i^{(m)}(t_1), b_i^{(m)}(t_1)$ $(1 \le i \le N - 1), \alpha_i^{(m)}(t_1), \beta_i^{(m)}(t_1), \gamma_i^{(m)}(t_1)$ and the matrix

$$A_1^{(m)} = A^{(m)}(t_1),$$

and the vector $\vec{\delta}_1^{(m)} = \vec{\delta}^{(m)}(t_1) = \left(\delta_1^{(m)}(t_1), \cdots, \delta_{N-1}^{(m)}(t_1) \right)^T$.

(ii) Then, the vector $X_2^{(m)} = \left(u_1^{(m)}(t_2), \cdots, u_{N-1}^{(m)}(t_2) \right)^T$ can be given by the following formula

$$X_2^{(m)} = \left[2E - (\Delta t)^2 A_1^{(m)} \right] (X_0 + \Delta t X_1) - X_0 + (\Delta t)^2 \vec{\delta}_1^{(m)}. \qquad (40)$$

C3. The vector $X_3^{(m)} = \left(u_1^{(m)}(t_3), \cdots, u_{N-1}^{(m)}(t_3) \right)^T$ is computed as follows.

(i) Computing the values $\hat{a}_i^{(m)}(t_2), (2 \le i \le N - 2), \bar{a}_i^{(m)}(t_2), b_i^{(m)}(t_1) (1 \le i \le N - 1), \alpha_i^{(m)}(t_2), \beta_i^{(m)}(t_2), \gamma_i^{(m)}(t_2)$ and the matrix

$$A_2^{(m)} = A^{(m)}(t_2),$$

and the vector

$$\vec{\delta}_2^{(m)} = \vec{\delta}^{(m)}(t_2) = \left(\delta_1^{(m)}(t_2), \cdots, \delta_{N-1}^{(m)}(t_2) \right)^T.$$

(ii) Then, the vector $X_3^{(m)} = \left(u_1^{(m)}(t_3), \cdots, u_{N-1}^{(m)}(t_3) \right)^T$ is defined by

$$X_3^{(m)} = \left[2E - (\Delta t)^2 A_2^{(m)} \right] X_2^{(m)} - (X_0 + \Delta t X_1) + (\Delta t)^2 \vec{\delta}_2^{(m)}. \qquad (41)$$

C4. The vector $X_{j+1}^{(m)} = \left(u_1^{(m)}(t_{j+1}), \cdots, u_{N-1}^{(m)}(t_{j+1}) \right)^T$ is computed as follows.

Suppose that we have calculated the vectors $X_1^{(m)}, X_2^{(m)}, \cdots, X_j^{(m)}$, then the vector $X_{j+1}^{(m)} = \left(u_1^{(m)}(t_{j+1}), \cdots, u_{N-1}^{(m)}(t_{j+1}) \right)^T$ can be determined by recurrence as follows

(i) Computing the values $\hat{a}_i^{(m)}(t_j), (2 \le i \le N - 2), \bar{a}_i^{(m)}(t_j), b_i^{(m)}(t_j) (1 \le i \le N - 1), \alpha_i^{(m)}(t_j), \beta_i^{(m)}(t_j), \gamma_i^{(m)}(t_j)$ and the matrix

$$A_j^{(m)} = A^{(m)}(t_j),$$

and the vector

$$\vec{\delta}_j^{(m)} = \vec{\delta}^{(m)}(t_j) = \left(\delta_1^{(m)}(t_j), \cdots, \delta_{N-1}^{(m)}(t_j) \right)^T.$$

(ii) Then, the vector $X_{j+1}^{(m)} = \left(u_1^{(m)}(t_{j+1}), \cdots, u_{N-1}^{(m)}(t_{j+1}) \right)^T$ can be determined by the following formula

$$X_{j+1}^{(m)} = \left[2E - (\Delta t)^2 A_j^{(m)} \right] X_j^{(m)} - X_{j-1}^{(m)} + (\Delta t)^2 \vec{\delta}_j^{(m)}. \qquad (42)$$

We do the process of computation until $j = M - 1$, then we can determine $X_j^{(m)}$ by

$$X_j^{(m)} = \left(u^{(m)}(x_1, t_j), \cdots, u^{(m)}(x_{N-1}, t_j) \right)^T, \ 1 \le j \le M. \qquad (43)$$

C5. The error of two iterative steps, the mth step and the $(m-1)^{\text{th}}$ step, is defined by

$$\left\| u^{(m)} - u^{(m-1)} \right\|_{M,N} = \max_{1 \le j \le M} \max_{1 \le i \le N-1} \left| u^{(m)}(x_i, t_j) - u^{(m-1)}(x_i, t_j) \right|. \qquad (44)$$

The iterative process will be stopped at the mth step when the error of two steps is satisfied

$$\left\| u^{(m)} - u^{(m-1)} \right\|_{M,N} < 10^{-4}. \qquad (45)$$

C6. The error of the approximate solution $u^{(m)}(x, t)$ and the exact solution $u_{ex}(x, t)$ is defined by

$$E_{M,N} = \left\| u^{(m)} - u_{ex} \right\|_{M,N} = \max_{1 \le j \le M} \max_{1 \le i \le N-1} \left| u^{(m)}(x_i, t_j) - u_{ex}(x_i, t_j) \right|. \qquad (46)$$

All computations were carried out using MATLAB software.

With $T = 0.1$, $N = 10, 20, 30, 40, 50$ and $M = N^2$, Table 1 describes the errors $E_{N,M}$ between the approximate solution $u^{(m)}(x, t)$ and the exact solution $u_{ex}(x, t)$, corresponding to the different meshes. The following figures describe the curved surfaces of approximate solution $u^{(m)}(x, t)$ corresponding to mesh of $N = 50, M = 2500$, and the curved surface of exact solution $u_{ex}(x, t)$ corresponding to the mesh of $N = 50$ and $M = 2500$. Specifically, we have some remarks as follows.

(i) At the stop of iteration process, Table 1 presents the errors $E_{N,M}$ between the approximate solution $u^{(m)}(x, t)$ and the exact solution $u_{ex}(x, t)$, corresponding to the different meshes. It is clear that the errors will be decreasing when the fineness of meshes is increasing.

Table 1. Errors of approximate solution $u^{(m)}(x,t)$ and exact solution $u_{ex}(x,t)$ corresponding to different meshes

N	M	$E_{N,M}$
10	100	0,004304964136035
20	400	0,003423100759947
30	900	0,002560797440241
40	1600	0,001966395627641
50	2500	0,001602844628803

(ii) Fig. 1 and Fig. 2, respectively, describe the curved surface of exact solution $u_{ex}(x,t)$ and of approximate solution $u^{(m)}(x,t)$ corresponding to mesh of $N = 50$ and $M = 2500$.

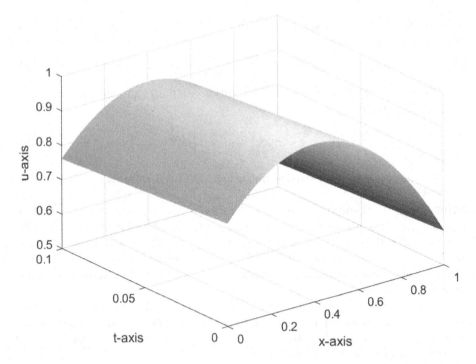

Fig. 1. The curved surface of exact solution $u_{ex}(x,t)$ corresponding to mesh of $N = 50$, $M = 2500$

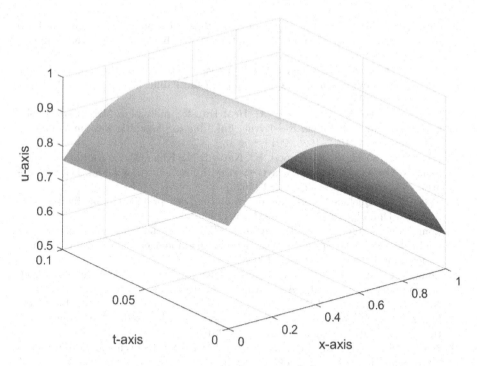

Fig. 2. The curved surface of approximate solution $u^{(m)}(x,t)$ corresponding to mesh of $N = 50$, $M = 2500$

4   Conclusions

In the present paper, we consider an initial-boundary value problem for a non-linear wave equation in which the source contains two boundary values. First, the existence of a recurrent sequence via the 2-order iterative scheme and its convergence to the unique weak solution of the proposed model are directly claimed by the results given in [12]. Next, a numerical algorithm for finding the approximate solutions corresponding to this scheme is constructed by the finite-difference method. To close the paper, a concrete example is numerically considered to evaluate the errors between the approximate solution and the exact solution.

Acknowledgements. The authors would like to thank the handling editor and the referees for the valuable comments and corrections for the improvement of the paper.

References

1. Agarwal, R.P.: Boundary Value Problems for Higher Order Differential Equations. World Scientific, Singapore (1986)

2. Gupta, C.: Solvability of a three - point nonlinear boundary value problem for a second order ordinary differential equation. J. Math. Anal. Appl. **168**, 540–551 (1992)

3. Il'in, V., Moiseev, E.: Nonlocal boundary value problem of the first kind for a Sturm-Liouvill operator in its differential and finite difference aspects. Diff. Eqns. **23**, 803–810 (1987)

4. Ijaz, N., Bhatti, M.M., Zeeshan, A.: Heat transfer analysis in MHD flow of solid particles in non-Newtonian Ree-Eyring fluid due to peristaltic wave in a channel. Thermal Sci. **23**, 1017–1026 (2019)

5. Iqbal, S.A., Sajid, M., Mahmood, K., Naveed, M., Khan, M.Y.: An iterative approach to viscoelastic boundary layer flows with heat source/sink and thermal radiation. Thermal Sci. **24**, 1275–1284 (2020)

6. Lions, J.L.: Quelques méthodes de résolution des problèmes aux limites non-linéaires. Dunod; Gauthier-Villars, Paris (1969)

7. Long, N.T., Dinh, A.P.N., Truong, L.X.: Existence and decay of solutions of a nonlinear viscoelastic problem with a mixed nonhomogeneous condition. Numer. Func. Anal. Opt. **29**, 1363–1393 (2008)

8. Long, N.T., Ngoc, L.T.P.: On a nonlinear wave equation with boundary conditions of two-point type. J. Math. Anal. Appl. **385**, 1070–1093 (2012)

9. Ngoc, L.T.P., Truong, L.X., Long, N.T.: An N-order iterative scheme for a nonlinear Kirchhoff-Carrier wave equation associated with mixed homogeneous conditions. Acta Math. Vietnam **35**, 207–227 (2010)

10. Ngoc, L.T.P., Tri, B.M., Long, N.T.: An N-order iterative scheme for a nonlinear wave equation containing a nonlocal term. Filomat **31**, 1755–1767 (2017)

11. Ngoc, L.T.P., Triet, N.A., Dinh, A.P.N., Long, N.T.: Existence and exponential decay of solutions for a wave equation with integral nonlocal boundary conditions of memory type. Numer. Func. Anal. Opt. **38**, 1173–1207 (2017)

12. Nhan, N.H., Ngoc, L.T.P., Long, N.T.: A N-order iterative scheme for the Robin problem for a nonlinear wave equation with the source term containing the unknown boundary values. Non. Func. Anal. Appl. **22**, 573–594 (2017)

13. Parida, P.K., Gupta, D.K.: Recurrence relations for a Newton-like method in Banach spaces. J. Comput. Appl. Math. **206**, 873–887 (2007)

14. Pinder, G.F.: Numerical Methods for Solving Partial Differential Equations. Wiley, New York (2018)

15. Triet, N.A., Ngoc, L.T.P., Dinh, A.P.N., Long, N.T.: Existence and exponential decay for a nonlinear wave equation with nonlocal boundary conditions of 2N-point type. Math. Meth. Appl. Sci. **44**, 668–692 (2021)

16. Truong, L.X., Ngoc, L.T.P., Long, N.T.: The n-order iterative schemes for a nonlinear Kirchhoff-Carrier wave equation associated with the mixed inhomogeneous conditions. Appl. Math. Comput. **215**, 1908–1925 (2009)

17. Truong, L.X., Ngoc, L.T.P., Dinh, A.P.N., Long, N.T.: Existence, blow-up and exponential decay estimates for a nonlinear wave equation with boundary conditions of two-point type. Nonlinear Anal. TMA **74**, 6933–6949 (2011)

Author Index

Printed in the United States
by Baker & Taylor Publisher Services